Faith in Fallacy

Faith in Fallacy

A Century of State-Sanctioned Science Denial

JAMES LAWRENCE POWELL

OXFORD
UNIVERSITY PRESS

Oxford University Press is a department of the University of Oxford.
It furthers the University's objective of excellence in research, scholarship,
and education by publishing worldwide. Oxford is a registered trade mark of
Oxford University Press in the UK and in certain other countries.

Published in the United States of America by Oxford University Press
198 Madison Avenue, New York, NY 10016, United States of America.

© James Lawrence Powell 2024

All rights reserved. No part of this publication may be reproduced, stored in
a retrieval system, or transmitted, in any form or by any means, without the prior
permission in writing of Oxford University Press, or as expressly permitted
by law, by license or under terms agreed with the appropriate reprographics
rights organization. Inquiries concerning reproduction outside the scope of the
above should be sent to the Rights Department, Oxford University Press, at the
address above.

You must not circulate this work in any other form
and you must impose this same condition on any acquirer

CIP data is on file at the Library of Congress

ISBN 9780197784686

DOI: 10.1093/9780197784716.001.0001

Printed by Integrated Books International, United States of America

To Joan, always.

Science is what we as human beings do best, at the end of the second millennium. Science is our equivalent of painting in Michelangelo's day, of music in the time of Bach, of seafaring in the age of Prince Henry the Navigator.
—Timothy Ferris, 1990[1]

Brief Contents

1. Humanity's Debt to Science   1

PART I   STATE SCIENCE DENIAL UNDER TOTALITARIANISM

2. Lysenko and the Origins of Soviet Pseudoscience   9
3. Refashioning Heredity   21
4. Pseudoscience Defeats Science   32
5. Rise and Fall   43
6. Big Brother, Little Brother   57
7. Jewish Physics   76
8. House of Shutters   90

PART II   STATE-SANCTIONED SCIENCE DENIAL IN DEMOCRACIES

9. AIDS   107
10. A Predictable Emergency   121
11. Protective Measures   132
12. Politicization of COVID-19 Denial   143
13. Global Warming: The Ultimate Triumph of Science Denial?   154
14. Roadmap to Doomsday   167

Acknowledgments   171
Notes   173
Index   189

Detailed Contents

1. Humanity's Debt to Science   1

   PART I STATE SCIENCE DENIAL UNDER TOTALITARIANISM

2. Lysenko and the Origins of Soviet Pseudoscience   9
   Pioneers of Science   9
   The Barefoot Professor   11
   Vavilov: The Mendeleev of Biology   15
   The 1929 Leningrad Conference   17
   Wreckers   19

3. Refashioning Heredity   21
   Statesman of Science   21
   Holodomor   23
   The Terror Begins   25
   Michurinism: Lysenkoism Rebranded   26
   Naming Names   28

4. Pseudoscience Defeats Science   32
   False Claims and Failed Experiments   32
   Gathering Clouds   35
   We Cannot Go on This Way   37
   Lysenko Towers Above   40
   The Death of Vavilov   41

5. Rise and Fall   43
   Sunflowers into Strangleweed   43
   The 1948 Lenin Academy Conference   45
   Song of the Forests   49
   At Khrushchev's Side   51
   Graham Visits Lysenko   53
   Bloodlands   54

6. Big Brother, Little Brother 57
   Mao and the Victory of Communism 57
   Michurinism Comes to China 58
   The Qingdao Symposium 64
   Mao's Constitution 65
   Tombstone 72

7. Jewish Physics 76
   Mein Kampf 76
   German Physics 80
   The Nazification of a Physicist 81
   Heisenberg and *Das Schwarze Korps* 86

8. House of Shutters 90
   The Eugenics Movement 91
   Nazi Racial Hygiene 95
   Hadamar 99
   The Hadamar Trials 100

PART II STATE-SANCTIONED SCIENCE DENIAL IN DEMOCRACIES

9. AIDS 107
   An Exotic New Disease 107
   The 1984 Press Conference 108
   Denialist Roles 110
   The Villain and the Head of State 115

10. A Predictable Emergency 121
    Crimson Contagion 121
    "This Is a Flu" 124
    Stop the Spread 128
    Quackery 130

11. Protective Measures 132
    The Spanish Flu 132
    Testing 133
    Masks 134
    Vaccines 138
    Operation Warp Speed 140

12. Politicization of COVID-19 Denial   143
    Red States vs. Blue States   144
    Falling Confidence in Medicine   147
    Lost Window of Opportunity   148
    The Tropical Trump   151

13. Global Warming: The Ultimate Triumph of Science
    Denial?   154
    The Politicization of Global Warming   155
    Merchants of Doubt   162
    The Perfect Metaphor   164

14. Roadmap to Doomsday   167

Acknowledgments   171
Notes   173
Index   189

1
Humanity's Debt to Science

Science and medicine have brought seemingly miraculous improvements in both the length and quality of human life. For centuries before the Age of Enlightenment in the seventeenth and eighteenth centuries, human life expectancy hovered around thirty years. Even by 1900, it had risen only to thirty-two years. Since then, global life expectancy has more than doubled, to 73.4 years in 2019. Most of the increase has occurred since 1960, when global life expectancy was only fifty-one years. Even individuals born today in countries with the lowest life expectancy can expect to live an additional fifteen years, compared to the global average in 1900. This improvement is due primarily to a fall in child mortality, but life expectancy has risen in every age category. Better sanitation and hygiene are part of the reason, but advancements in medical science have been equally important. In 1900, there were no antibiotics, EKGs to detect heart problems, fetal ultrasound, kidney dialysis, pacemakers, sulfa drugs, and so on. There were no vaccines for diphtheria, hepatitis B, human papillomavirus (HPV), influenza, measles, mumps, rubella, polio, rabies, tetanus, yellow fever, and whooping cough. The World Health Organization (WHO) says that of all these medical inventions, vaccines have saved more human lives than any other.[1]

Since the beginning of the twentieth century, the *quality* of human life for many has also improved markedly, though disparities exist within and between nations. At least in advanced countries, people today on average have higher work productivity and more leisure time, higher levels of education and living standards, and higher incomes (adjusted for inflation). Pivotal twentieth-century inventions are responsible for much of this improvement. A list of those that have had the greatest impact include the airplane, automobile, computer, electric refrigeration, electronics, paved highways, household appliances, the internet, lasers, plastics, radar, radio, rural electrification, telephone, television, transistor, and wireless technology.

Every educated person knows the source of these improvements: science. Without it, human life would resemble that of the Middle Ages: nasty, brutish, and short, in the words of philosopher Thomas Hobbes.[2] Given this record, one would think that the first rule for heads of state and their governments would be to follow where the best science leads. But as we will explore in this book, the clear lesson from history is that when science comes up against ideology and ignorance, it often loses. When that happens, the assumption seems to be that a person or government can choose which parts of science to accept and which to reject—and pay no price. But science is like an interwoven tapestry in which each thread supports all the others, strengthening the entire fabric. One cannot logically reject the findings of one branch of science—pull one thread—because it happens not to fit one's ideology, while accepting all the others. But many abandon logic and do exactly that.

Science is a systematic and logical approach to discovering how the universe works, based on empirical evidence and the testing and refining of hypotheses. Systematic because science follows a proven methodology for getting at the truth. Logical because scientists work with effects and use reason to discover causes. Empirical evidence is that produced by experiment or observation, which scientists then devise hypotheses to explain. They test and revise, choosing the hypothesis that best explains the evidence. As more evidence is gathered and tested, a hypothesis grows stronger and can be promoted to the status of theory.

A critical development in the history of science was the introduction in the seventeenth century of journals in which scientists could share their results. The first in English was the *Philosophical Transactions of the Royal Society*, which began publication on March 6, 1665. Scientists quickly embraced journal publication not only to establish their priority in discovery but also to describe their methodology so that others could replicate their work, thus confirming its reliability. Journal publication disseminated knowledge widely and launched a burst of discovery that continues to this day.

Science denial, in contrast, is the rejection of settled science despite its endorsement by a broad consensus among scientists.[3] It is motivated by some combination of ideology, politics, personal beliefs, and vested interests. One distinguishing characteristic is that because science denial rests on a rejection of empirical evidence, new facts and discoveries almost never persuade science deniers to change their minds. For them, ideology always trumps facts, and often they carry their denial to the grave.

Science deniers sometimes simply declare a theory false, without attempting to replace it with one of their own. Manmade global warming is an example: those who deny it have no other theory to explain why, as fossil fuel emissions and atmospheric carbon dioxide (CO_2) have risen, global temperature has climbed in lockstep. In these cases, science deniers would rather have no theory than one that they find inconvenient or that violates their ideology.

In many instances, however, science deniers adopt an alternative explanation that amounts to pseudoscience: a claim for which there is no empirical evidence, was not reached via a rigorous methodology, is widely rejected by the scientific community, and has never been endorsed in a peer-reviewed scientific article. The best-known example is biblical creationism: the belief that a divine being created the universe and all living organisms.[4] Those who adopt it reject Darwin's widely accepted theory of evolution, even though it is supported by overwhelming scientific evidence. Other examples of pseudoscience include astrology, crystal healing, dowsing (using a stick or medal rod to detect underground water), the flat earth, homeopathy (the claim that highly diluted materials can heal), folk medicine, and anti-vaccination.

One example of pseudoscience that will loom large in this book is the belief that traits acquired during life can be passed on to subsequent generations, known as the *inheritance of acquired characteristics*. An alleged example is the giraffe, whose necks are claimed to have elongated over generations as they stretch to reach leaves high in trees. They then pass on this acquired trait to their offspring. But it is genes, and not traits gained in life, that are inherited.

Even though disproven by genetics, inheritance of acquired characteristics became the basis for state science policy in the USSR and the People's Republic of China. These are the first two cases we will take up. They are examples of how state science policy can be imposed top-down by an all-powerful dictator. In the USSR, though Stalin believed in the inheritance of acquired characteristics for all his life, he did not impose it directly on Soviet biology, but rather through a convenient agent named Trofim Lysenko, whom he praised publicly and anointed as a Hero of the Soviet Union. Mao's Red China would adopt Lysenkoism in its entirety, with catastrophic effects. Lysenkoism in the USSR has produced an entire field of scholarship and a voluminous literature, making it the canonical example of state science denial. We have far more information about Lysenkoism than any of the other historical examples, so it is appropriate to give it detailed attention in this book.

Adolf Hitler made state policy of Nazi eugenics, a pseudoscientific concept which held that some races have more intelligence and other desirable traits than inferior ones. Germans were allegedly descended from a master race (*Herrenrasse*), the Aryans, an obsolete group of Indo-Europeans who migrated into the Indian subcontinent. Aryan supremacy justified the sterilization, and eventually the murder, of those deemed undesirable. Typical Aryan traits were supposedly fair skin, light hair color, and blue eyes. These features were glorified as the epitome of beauty and racial superiority, despite their absence among the top Nazis. Hitler imposed a state policy that denied Jews positions in universities and research institutes, which led many to flee Germany for Allied nations, whose war effort they aided. The Holocaust had its beginning in Nazi eugenics.

But science denial and pseudoscience can become state policy not only in cruel dictatorships but also in democracies. In South Africa, President Thabo Mbeki conducted his own internet research and concluded that HIV did not cause AIDS and that folk remedies were preferable to antiretroviral drugs. Through delay and inaction, and by those he appointed as ministers, he made AIDS denial and pseudoscientific therapies the effective policy of South Africa, at the cost of hundreds of thousands of lives.

Brazil and the United States provide additional examples of a policy of science denial set by democratically elected leaders. Both President Jair Bolsonaro and President Donald Trump downplayed the danger of the COVID-19 virus and discouraged protective measures. Both called it nothing more than a flu. Trump deserves credit for promoting the rapid development of vaccines against the virus, but large numbers of Americans ignored his advice and adopted anti-vax pseudoscience. Eventually, he joined them. In this case, it was the people who in effect set a state policy of science denial and the leader who followed.

The rejection of manmade global warming, either directly or through inaction and delay, threatens more deaths than all other examples of science denial put together. Those who deny it, which includes nearly every elected Republican, are betting their grandchildren's future on the transparently false belief that the world community of scientists is wrong about a matter of science and that they are right.

With global warming, the ultimate cost of state science denial has become frighteningly clear. Nations and their leaders have a choice: either learn from the examples we review in this book, trust scientists and act on their advice, or cripple the lives of coming generations.

PART I
STATE SCIENCE DENIAL UNDER TOTALITARIANISM

PART 1

2
Lysenko and the Origins of Soviet Pseudoscience

"He smiled only once, this barefoot scientist."[1]

—*Pravda*

Pioneers of Science

In the years before the 1917 Revolution, Russian biologists were among the world's foremost researchers. The most celebrated was Ivan Pavlov (1849–1936), renowned for his behavioral experiments with dogs. He discovered that the animals began to salivate as soon as the person who regularly fed them appeared, even before any food was offered. By systematically sounding a buzzer just before delivering food, Pavlov observed that the noise alone triggered salivation. This phenomenon became known as the *conditioned reflex*, a discovery so significant that it earned Pavlov the 1904 Nobel Prize in Physiology or Medicine. Another distinguished Russian scientist in the early twentieth century was Nikolai Koltsov (1872–1940). Before his research, biologists had postulated that a cell's internal osmotic pressure controlled its shape. Koltsov demonstrated that instead an assembly of fibers or tubules forms a skeletal framework to which the cell conforms. In 1927, he proposed that the inheritance of characteristics across generations is controlled by a "giant hereditary molecule" composed of "two mirror strands that would replicate in a semi-conservative fashion using each strand as a template."[2] In 1953, James Watson and Francis Crick, who evidently had not known of Koltsov's hypothesis, announced their discovery of the famous double-helix structure of the DNA molecule and won the 1962 Nobel Prize in Physiology or Medicine. In 1917, only months prior to the Russian Revolution, the Institute of

Experimental Biology, named for Koltsov, opened in Moscow. The institute's Laboratory of Genetics was headed by Sergei Chetverikov (1880–1959), who went on to establish the science of population genetics and to explain the critical role of recessive mutations in heredity. Chetverikov was the first Russian to work with populations of the *Drosophila* fruit fly, which were so critical to advances in genetics in the West. His work showed that Darwinian evolution and genetics were not in conflict.

A few months before the Russian Revolution began, Vladimir Lenin wrote a book titled *The State and the Revolution*, in which he explained that in the coming Communist state simple workers and poor peasants would control the work of intellectuals.[3] Some 2,000 of the most outspoken opponents of his policy were banished from the new Soviet Union and the remainder placed under the control of commissars. Pavlov, however, was the exception. A favorite of Lenin, he was able to get away with criticizing Soviet Communism when others could not. In 1923, for example, he said publicly that he would not sacrifice the hind leg of a frog to the type of social experiment that the Soviet regime was conducting. In a 1927 letter, Pavlov told Stalin, now general secretary of the Communist Party, that "On account of what you are doing to the Russian intelligentsia—demoralizing, annihilating, depraving them—I am ashamed to be called a Russian!"[4] He later complained to Vyacheslav Molotov, Stalin's foreign minister, about the mass persecution of intellectuals.[5] Pavlov's fame allowed him to escape Stalin's retribution and die a natural death in Leningrad in 1936. The year before his death, to express the debt that humanity owed to his animals, Pavlov had a bronze sculpture of an unnamed dog placed on an inscribed pedestal in the courtyard of the Institute of Experimental Medicine in Leningrad. It still stands.

Koltsov and Chetverikov were not as fortunate. Their fate foreshadowed that of countless Soviet researchers. In 1920, Koltsov aligned himself with a rebellious group of intellectuals, all of whom were arrested. He was sentenced to death, but his friend, the writer Maxim Gorky, interceded with Lenin on his behalf, leading to Koltsov's liberation and reinstatement as director of the Institute of Experimental Biology. However, two decades later, amid Stalin's terror, the NKVD (the Soviet secret police) summoned Koltsov to testify against another esteemed Russian scientist, Nikolai Vavilov. Following a ruthless interrogation, the NKVD poisoned Koltsov.[6] As for Chetverikov, he was arrested in 1929 on trumped-up charges, exiled for five years, and had his laboratory destroyed. On his release, Chetverikov founded the Department of Genetics at Gorky University. In 1948, he was denounced for the crime of supporting modern genetics, which the Politburo had banned, and fired. He died forgotten and unrecognized for his seminal contribution to biology.

These great scientists were primarily theorists and researchers. But what the new Soviet state faced was the practical necessity of feeding its people. For that, it needed grain, especially wheat, which had been Russia's chief staple crop and one of its most important exports. The Revolution and subsequent civil war had severely disrupted agricultural production in the country and crippled the railway system, necessary to transport grain. Reflecting the Soviet goal of a classless society, large landowners and wealthy farmers would soon have their land confiscated and turned over to collectives, causing a significant decline in agricultural output. All this put the emphasis on soil management and crop production (the science of agronomy) rather than on theoretical biology.

The Barefoot Professor

In August 1927, an article titled "The Fields in Winter" appeared in the Soviet newspaper *Pravda* (Truth), the official organ of the Communist Party. Written by a well-respected journalist, the story had an engaging style and a memorable description of its subject, a young agronomist named Trofim Denisovich Lysenko. According to the article, the twenty-nine-year-old "barefoot professor" had, working in remote Azerbaijan, "solved the problem of fertilizing the fields without fertilizer and minerals." He had grown a winter crop of peas that "turn[ed] the barren fields of the Transcaucasus green," providing fodder so that cattle would not go hungry during the frigid winters and fertilizing the soil through the nitrogen fixation of the pea plants. "Of doleful appearance ... stingy with a word ... he smiled only once, this barefoot scientist."[7] Figure 2.1

12 FAITH IN FALLACY

Figure 2.1 Trofim Denisovitch Lysenko
Wikimedia Commons

Lysenko's remarkable crop of winter peas was evidently never attempted again—or if it was, failed. Instead, subsequent articles would favorably describe his experiments with a process called *jarovization* in Russian, translated as *vernalization* in English, from the Latin *vernum* for spring.[8] In this method, seeds of wheat and other grains are cooled and moistened during their germination, causing them to flower sooner when planted and to produce higher yields. Farmers had long known that some plants require exposure to cold to flower—provided they do not freeze. During the winter of 1927–1928, for example, unusually cold weather in Ukraine drastically reduced winter wheat production. This set the stage for an even more devastating famine (known as the *Holodomor*) that would arrive a few years later. If winter wheat (sown in autumn or winter for harvesting the following summer) manages to survive the cold months, however, it produces higher yields than spring wheat (sown in spring for harvesting the following autumn). Lysenko reported that he had soaked winter wheat seeds in water, cooled them, and then planted them in late winter or early spring after the worst of the cold weather was over. This was nothing new, as vernalization had long been known to growers. But then, Lysenko claimed, once winter wheat had been vernalized, it would have been transformed *permanently* into spring wheat and would not have to be vernalized again. But this appealed to an outmoded theory of evolution called the *inheritance of acquired characteristics.*

Decades before Darwin published his theory of descent with modification, another theory of heredity had prevailed. In 1809, French scientist Jean Baptiste Lamarck (1744–1829) postulated that offspring can inherit traits their parents acquired during their lifetimes, using as one example the giraffe. Lamarck claimed that as these lofty creatures stretch to reach enticing foliage, their necks incrementally elongate, a characteristic they transmit to their descendants. Eventually, all giraffes have long necks. As another example, Lamarck cited the blacksmith, whose progeny would purportedly inherit his robust muscles and bequeath them to future generations. The inheritance of acquired characteristics was widely accepted throughout the nineteenth century and even embraced by Darwin, who, unaware of genes and the rules of their inheritance, believed that heightened utilization of a trait would increase its likelihood of inheritance. This idea, which Darwin dubbed *pangenesis*, had ancient roots, going back to the Greek scholar Hippocrates. Darwin theorized that cells in disparate body regions release tiny particles, or gemmules, that converge in eggs and sperm, from which they transfer to offspring. This concept became known as Darwin's *soft inheritance.*[9]

Farmers and animal breeders had long recognized that systematic crossbreeding could promote certain traits considered desirable in offspring, albeit without guarantee. The process by which this happens was discovered by Czech scientist and abbot Gregor Mendel (1822–1884) in his investigations of pea plants during the 1850s and 1860s. He studied various pea attributes—plant height, pod shape, seed shape and color, flower position, and so on—and their progression in subsequent generations. Mendel discovered that when a plant with yellow seeds was crossbred with one bearing green seeds, their immediate progeny produced only yellow seeds. The yellow color exemplified what would later be termed a *dominant* trait. In the following generation, however, there was always a single green seed for every three yellow seeds. The green pigment was *recessive*: concealed but not eradicated, as Chetverikov would discover. These imperceptible *factors*, as Mendel labeled them, and not traits they acquire during their lifetimes, are what organisms pass on to the next generation.

As has sometimes happened with great theories in the history of science, Mendel published in an obscure journal, leaving his findings to be rediscovered independently. Then in 1905, British biologist William Bateson (1861–1926) established the field of genetics. Later in the twentieth century, a series of biologists would amalgamate natural selection, Mendelian genetics, and population genetics into a mathematical theory dubbed the *modern synthesis*, a term popularized by Julian Huxley in his 1942 book *Evolution: The Modern Synthesis*.[10]

The discovery of genes debunked both Darwin's soft inheritance and the inheritance of acquired traits, yet Lamarck's hypothesis persisted. In a sense, it seemed to embody a kind of intuitive logic: we share numerous similarities with our parents, so why not in attributes they developed during their lifetimes and passed on to us? We can observe those traits; microscopic genes we cannot. This may help explain why Josef Stalin and many Soviet scientists accepted the inheritance of acquired characteristics, to the point that within one generation it became the official science policy of the USSR.

Lysenko was born in 1898 to a peasant family in Ukraine and did not read until age thirteen. He attended a vocational school for gardeners and participated in a one-month course on growing sugar beets, an important crop in the new USSR during the early twentieth century. Lysenko went on to graduate from a horticultural school in 1921, and in 1925 he received a doctorate from the Kiev Agricultural Institute (now the National University of Life and Environmental Sciences of Ukraine). His first job was as junior agronomist at the Gyandzha Experimental Agricultural Station in Azerbaijan, near the Caspian Sea.[11] Lysenko's claim that he could permanently change one species into another fit with Soviet ideology and was accepted without ever being tested scientifically. Lysenko's specious claim would make him the most acclaimed scientist in the USSR. Statues were erected in his honor, and when he gave a lecture, a band would play music, and songs of praise would be sung. Lysenko won the Stalin Prize in science and engineering, was named a Hero of Socialist Labor, and received the highest Soviet decoration—the Order of Lenin—seven times.

The year 1929 was a momentous one in the Soviet Union, as Stalin announced that it would be the year of the "Great Break" with the past, aimed at rapidly transforming the USSR from a primarily agrarian society to an industrialized and militarized superpower. A new economic policy would be put in place, emphasizing heavy industries like steel, coal, and machinery, and more factories, dams, and railways would be built. To increase agricultural efficiency and allow the state more resources to invest in industrialization, small-scale, privately owned farms were combined into large, state-controlled collectives known as *kolkhozes*. The *kulaks* (farmers affluent enough to possess their own farm and employ laborers—in other words, the most accomplished) were exterminated. Others who defied the regime were banished to the gulags. Crop failures ensued, and dispossessed, resentful peasants wandered the countryside. Not only agriculture, but Soviet culture as well, was to undergo revolution under the Great Break. Suspect "bourgeois specialists": scientists, engineers, and doctors who came from wealthier families and had been trained before the Revolution had been tolerated out of necessity, but soon a new generation of students schooled in Soviet ideology would be ready to replace them.

Vavilov: The Mendeleev of Biology

On August 7, 1927, came the article in *Pravda* touting the "barefoot professor" who did not fiddle with "the hairy legs of flies (referring to scientists like Chetverikov and American biologist Thomas Morgan, who were exploring genetics using fruit flies) but went to the root of things," turning the winter fields green with peas for forage and compost.[12] This was likely when Nikolai Vavilov first heard of Lysenko. As we will see, Vavilov looms large in the story of science denial at the level of the state, personifying its deadly cost. Born to a wealthy family in Moscow in 1887, Vavilov's background was the opposite of Lysenko's. In 1906, he entered Moscow's Petrovskaya Agricultural Academy, known as the *Petrovka*. After earning his degree, he studied genetics with Bateson at Cambridge. There Vavilov researched the resistance of wheat to fungi, which would become a career-long interest. On his return, he served from 1917 to 1920 as professor of agronomy at the University of Saratov, a port on the Volga River upstream (north) of Volgograd. Twenty years later, under vastly different circumstances, Vavilov would return to Saratov. In 1924, Vladimir Lenin personally appointed him as head of Applied Botany at the Lenin All-Union Academy of Agricultural Science, which I will refer to as the *Lenin Academy*. It comprised a large, Union-wide network of research institutions. Vavilov would serve as its director from 1929 to 1935. In our discussion of science denial in the USSR, we will follow the careers of the two rivals, Lysenko and Vavilov, and their opposite outcomes.

From his eminent position at the Lenin Academy, Vavilov became an internationally acclaimed plant biologist and an enthralling lecturer. Colleagues hailed him as the Mendeleev of biology, referring to the Russian chemist who had invented the periodic table of chemical elements. Vavilov conducted dozens of trips abroad to collect plants, and he spoke the major European languages and picked up a smattering of others as necessary in his travels. He was said to sleep only four or five hours a night and to do his writing while traveling. He won numerous awards and served as vice president of the Sixth International Congress of Genetics, held at Cornell University in 1932. Vavilov had been elected president of the Seventh Congress, to be held in February 1937 in Moscow. But Stalin canceled the meeting, and it was moved to Edinburgh in 1938. But by then, Soviet scientists were forbidden to travel abroad. On September 1, 1939, Nazi Germany invaded Poland, launching World War II.

Vavilov's best-known scientific contribution was his hypothesis that a given plant species likely originated where we now find its highest diversity: this is its *center of origin*. He identified a number of primary centers, including Southwest Asia, Central Asia, South and East Asia, the Mediterranean region, Abyssinia, South Mexico and Central America, the Andes, and the southeastern United States. Plants collected from these centers likely retained their original gene pool and therefore would be the most useful in crossbreeding.

When he first learned of Lysenko, Vavilov—open-minded to a fault—sent a member of his staff to investigate. The report was not flattering, describing Lysenko as "an uneducated and extremely egotistical person, deeming himself to be a new Messiah of biological science."[13] But Vavilov was attracted by Lysenko's evident dedication and hard work and planned to invite him to visit Vavilov's experimental station outside Leningrad. However, the institute's plant specialist, Nikolai Maksimov (1880–1952), objected, mainly on the grounds that Lysenko did not know a foreign language and could not read the international scientific literature and thus had little in the way of science to contribute. Vavilov deferred to Maksimov's judgment and did not invite Lysenko. We might imagine that this was a missed opportunity to establish a colleagueship that would have prevented the coming enmity between Lysenko and Vavilov, but it would turn out that Lysenko did not have colleagues, only followers and enemies.

The 1929 Leningrad Conference

In 1929, Vavilov was elected a full member of the Academy of Sciences of the USSR and confirmed as president of the Lenin Academy by the Council of People's Commissars. He organized the first Soviet Congress on genetics and plant breeding, to be held in Leningrad in January 1929. Vavilov was also one of the most active and successful Soviet biologists, having earned an international reputation. To many abroad, his was the face of Soviet science. Vavilov had by this time collected a large repository of seeds from around the world and thought that vernalization might be used to cause different plants to flower at the same time, which would help in his crossbreeding experiments. He was already somewhat dubious about Lysenko's methods, but in true scientific spirit was willing to provide him a platform at the conference.

Lysenko's claims got a cool reception, with Maksimov saying, "The results [on vernalization] obtained by Comrade Lysenko do not represent anything new in principle, are not a scientific discovery in the precise sense of the word."[14] Another plant specialist agreed: "From the point of view of plant physiology, Agronomist Lysenko has made no discovery."[15] This time *Pravda* failed to mention Lysenko in its article on the conference.

One might have thought that the rebuff of his report at the conference would have sent Lysenko back to the field and laboratory to test and refine his theories and perhaps develop new ones, in hopes of a better response the next time he presented. Instead, he responded with what was essentially a publicity stunt, setting him on a "collision course with the scientific establishment and traditional science."[16] Lysenko and his father soaked a sack of winter wheat seed and buried it in a snowbank, where it remained until spring. Father Denis Lysenko then took out the seed and sowed it beside a field of spring wheat. Winter wheat does not ordinarily ear (develop grain-bearing heads or spikes) in the spring, but vernalization worked, and his father's winter wheat produced a better harvest than the nearby field of spring wheat. As historian David Joravsky writes, "At least, that is what the Lysenkos claimed."[17] An investigating committee organized by the Ukrainian Commissariat of Agriculture visited the father's farm, wrote that it had confirmed the claim, and ordered a 1,000-hectare test (one hectare is 10,000 square meters or about 2.5 acres). Before the results of the tests had come in (indeed, even before they had begun), the Ukrainian Commissariat announced the sensational news to the papers. *Pravda* now followed suit, writing that "agronomist Lysenko's discovery will lead our agriculture onto a high road of vast possibilities and extraordinary achievements and greatly increase the tempo of our socialist construction."[18] Throughout his career, Lysenko would get such effusive encomiums for his announced plans, each of which came before the previous claim had been adequately tested, to be followed soon by the next claim, and so on. Few acknowledged that his promises seldom delivered.

Lysenko's report at the conference and the well-publicized event with his father made it appear that vernalization offered an escape from the failing wheat crops and resulting famines that threatened not only millions but the Revolution itself. The official agricultural newspaper put out by Y. A. Yakovlev (1896–1938), who had just been promoted to All-Union Commissar of Agriculture, endorsed vernalization and in November 1929 asked leading plant specialists their opinion of Lysenko's plan.[19] They responded with the typical reluctance of scientists to criticize each other in the press, a caution now exaggerated by the risk of appearing unwilling to try anything that might possibly ward off famine. While having their doubts about vernalization, Soviet scientists spoke approvingly of Lysenko's emphasis on practical solutions. As a result, Joravsky writes that these cautious scientists put themselves "into a box that would confine them for 35 years." Later they would retreat even further into the box, some simply grafting Lysenkoite language onto their previous writings.[20]

The claim that changes in environment could produce permanent heritable traits fit irresistibly with Marxist ideology. Plants could be permanently improved by the right environment and so could Soviet men and women—under Communism.

Wreckers

To ensure that government officials, rather than the traditionally independent scientists, would direct Soviet agriculture, Vavilov's Lenin Academy was made a *secretariat* overseen by a new Commissariat of Agriculture. Perhaps because he had no choice, Vavilov diplomatically accepted the new direction, saying that it merely represented "two aspects of the same drive toward more efficient production."[21] In early June 1929, he left on yet another plant-collecting trip, this one to China and Japan.

Another threatening development took place that summer when the Communist Party set up a special committee responsible for "cleansing" the Academy of Sciences of "class enemies" among its member scientists.[22] Ominously, the committee included only one scientist. Some 700 academicians were dismissed not because of anything they did, but merely because they came from noble or bourgeois families. As Joravsky sums up, "Any critical comment, any silence that could be interpreted as criticism or mental reservation, became grounds for dismissal, or jailing, even shooting."[23] One of the first victims among the agriculturalists was Alexei Doyarenko, Vavilov's one-time university professor, who was arrested and imprisoned for "sabotage," an increasingly common defamatory label. Doyarenko's crime had been his failure to join in a paean to the new Soviet agrarian plans. His wife had recently died, and a friend who had sung at her funeral was threatened with dismissal because this smacked of outlawed religion. The friend responded that he was merely fond of music, but that did not suffice, and he was exiled.[24] Doyarenko was lucky: he was released and lived to 1958. Out of the Academy faculty of 168, twenty were dismissed and five sent to jail or executed under the catch-all "wreckers."[25] Joravsky defines this term as a translation of the Russian *vrediteli*, a "deftly venomous epithet whose basic meaning is 'pests.'"[26] To be labeled a wrecker by a Soviet official or newspaper was to stand on the brink of a very slippery slope.

3
Refashioning Heredity

"Bravo, Comrade Lysenko, Bravo."[1]

—*Stalin*

Statesman of Science

Though Vavilov was unaware of it, he had come under the scrutiny of the Soviet secret police in 1921 when he traveled to America to visit breeder Luther Burbank and geneticist Thomas Morgan. An agent of the OGPU, the Russian secret service (to be succeeded by the NKVD), followed Vavilov wherever he went. Agent "Svezda" reported that during his visit, Vavilov had told Morgan that the Russian people could not wait for the return to capitalism. Vavilov, the agent said, was "an adventurist who put his own interests above the Soviet state and built glory for himself at the expense of others."[2] Other agents followed Vavilov on his many trips outside of the USSR. In 1930, the OGPU opened an operational file on Vavilov and began gathering evidence to show that he had attempted to sabotage food production. The Vavilov file would swell to seven volumes. If there is one rule of the police state, it is that secret agents must file reports; otherwise, they would soon be out of a job. If they find nothing, they make up something. In a move that Franz Kafka and George Orwell would have understood, in September 1930 the OGPU announced that it had discovered the existence of a counterrevolutionary political group called the *Peasant Labor Party* (TKP).[3] That the TKP did not exist did not prevent the arrest of thousands, many of whom (and years later Vavilov) would confess to membership in the nonexistent organization as the least of a bad set of alternatives. The arrests were aimed at agricultural experts and food-production officials, many of whom confessed to sabotage. After short trials and verdicts from which there was no appeal, some fifty agriculturalists were summarily executed. Some who confessed were released to become informers for the OGPU. One of them, Ivan Yakushkin, had known Vavilov since they were students. In September 1931, he accused Vavilov of organizing "wrecking activities" in plant collection and breeding. Like the secret agents, prisoners released as informers had to report something, invented if necessary, or find themselves back in prison, or worse. One scientist, Alexander Kol, who had once lost an argument with Vavilov about research policy, saw his chance to get even. He wrote an article for a government economic newspaper in which he labeled Vavilov a counterrevolutionary. A newspaper editorial endorsed Kol's claim. Standard practice would have been for Vavilov to issue an apologetic "self-criticism," but he proudly refused.[4] In 1930, he was off again to America, where the same agent who had tailed him in 1921 again picked up his trail.

On this trip Vavilov, a true Soviet patriot to the end, tried to persuade two Russian scientists working in Morgan's California lab to return home. One was Theodosius Dobzhansky, a geneticist who had emigrated to the United States in 1927. He declined Vavilov's invitation and became a professor at Cal Tech and then at Columbia. Dobzhansky is best known for his 1937 book *Genetics and the Origin of Species*, which brought together Darwin's natural selection and Mendelian genetics. The other was Georgy Karpechenko, who had been awarded a Rockefeller Prize for producing a fertile cross between a radish and cabbage, something that had been thought impossible since they are two different species.[5] Karpechenko agreed to return to head the genetics lab at Vavilov's institute. He was arrested by the NKVD, charged with belonging to the same anti-Soviet group as Vavilov, and executed on July 28, 1941.

Vavilov could not have failed to recognize the threat that Lysenko posed to Soviet agriculture and to himself. Yet in the best spirit of science, he supported Lysenko's practical work as a complement to his own genetic research. Vavilov made it clear that he did not believe that the labor-intensive vernalization could be scaled up to mass crop production, but the government agricultural administrators disagreed and ordered large-scale trials, instructing Vavilov to assist Lysenko. In response, Vavilov invited Lysenko to attend the August 1932

International Congress on Genetics to be held at Ithaca, New York. To this collegial overture, Lysenko never responded.[6]

After the congress, Vavilov set off on a collecting expedition to Central and South America. On his way home, he stopped in Paris to visit colleagues, one of whom happened to be already on file with the OGPU. When Vavilov embraced his friend on the train platform before departing for Berlin, an agent who had been tailing him immediately notified his bosses of this further evidence of Vavilov's unreliability—and another damaging document dropped into Vavilov's growing OGPU file. In Berlin, Vavilov met with American geneticist H. J. Muller, who was working at the Kaiser Wilhelm Institute. In 1946, Muller would win the Nobel Prize for his work on mutations in fruit flies. He was an example of those scientists whom a supporter of Lysenko would deride as "people haters and fly lovers."[7] Vavilov invited Muller to come to the Lenin Academy as director of genetics, and Muller accepted. This was a coup for Vavilov and Soviet science, but the OGPU saw instead a dangerous fraternization with a fly-loving foreigner, and another note fell into Vavilov's file.

Holodomor

After the Russian Revolution, famine was an omnipresent threat. Initially, Lenin and his comrades had ample cause to harbor hope for Soviet agriculture. Imperial Russia had been a formidable wheat exporter, with Russian wheat comprising over one-third of global exports by 1910. However, the situation swiftly unraveled. The turmoil incited by the Civil War, the war-era government strategy of seizing grain, and the Russian rail system's inadequacy in transporting sufficient grain precipitated a devastating famine in 1921–1922, claiming the lives of an estimated five million people. By the fall of 1931, another famine was underway. The government's solution was to demand that, within four years, agronomists were to develop new, certified, high-yield crop varieties of potatoes and other crops, rather than taking the ten to twelve years that the agronomists said would be necessary. For wheat, the most important crop, the target time was reduced to two years. And what would allow this miracle? *Collectivization*: the consolidation of individual landholdings and labor from small, privately owned farms into giant collective farms. This would, it was assured, achieve greater agricultural efficiency, increase food production, and free up labor for industrialization. It did not work. Vavilov pleaded for more time to develop an agricultural master plan, but was rebuffed in favor of Lysenko, who promised much faster results.

These attempts at planning came too late, as during 1931–1932, an additional five million people perished from starvation in the Soviet Union, approximately four-fifths of them Ukrainian, who refer to this era as the *Holodomor*, signifying "death by hunger." Though the collapse of collectivization had rendered the 1932 harvest subpar, it likely would have sufficed to prevent widespread starvation. Instead, the Communist Party imposed unattainable grain quotas that farmers had to fulfill before retaining enough for their own consumption. Squads of government enforcers descended upon Ukraine to confiscate grain, and homes underwent routine searches. A new law deemed theft of socialist property a capital offense, leading to the execution of peasants for pilfering a sack of their own wheat. During this time, Moscow exported over a million tons of grain to other Soviet republics and abroad, exacerbating Ukraine's famine. As Khrushchev would say in his 1956 denunciation of Stalin, "Their method was like this. They sold grain abroad, while in some regions people were swollen with hunger and even dying for lack of bread."[8] Cannibalism became rampant, and the Soviet government, rather than ending grain seizures, displayed signs proclaiming, "Devouring your own children is an act of barbarism."[9]

Rafael Lemkin (1900–1959), the Polish attorney who defined the term *genocide*, believed that it applied to the Holodomor. Ukrainians have not forgotten what the Soviet government did to their grandparents' generation. They understand the wisdom of Nobel laureate Amartya Sen's statement: "Starvation is the characteristic of some people not *having* enough to eat. It is not the characteristic of there not *being* enough to eat (italics in original)."[10] Russian peasants said it more simply, "God makes a poor harvest, but human beings make a famine."

The Terror Begins

In 1933, after his return from America, Vavilov, in his position as director of the Institute of Genetics of the Academy of Sciences, was called to account before the Council of People's Commissars, the government budgetary office.[11] They scolded him for his expensive and, in their eyes, unnecessary trips abroad and also for the evident failure of his work to improve Soviet agriculture, conveniently ignoring the fact that all his plans had been rejected. Vavilov offered to resign but was refused. Next, he was banned from foreign travel, even to attend international scientific conferences. The secret police began to arrest scientists on Vavilov's staff, including Maksimov, who had been among the first to criticize Lysenko at the 1929 conference.

Near the end of 1933, an OGPU official wrote to Stalin making accusations against Vavilov and urging his immediate arrest. But no action was taken. Perhaps Vavilov was able to escape the fate of his colleagues, at least for a while, because he was the face of Russian science internationally. Vavilov continued to extend the hand of friendship to Lysenko, inviting him to come to Leningrad and visit the institute's experimental farm. Again, Lysenko did not reply.

Vavilov continued trying to conciliate Lysenko, telling a British visitor that "young men like Lysenko who 'walked by faith and not by sight,' might discover something new."[12] He went so far as to recommend Lysenko for several scientific awards and nominated him to the Soviet Academy of Sciences. But criticism of Vavilov only grew more damning, both in secret and openly. Two prominent members of the party wrote a covert letter to Stalin, saying that Vavilov "defended wreckers" and, as president of the Lenin Academy, was "a negative entity" who preferred to go on long foreign trips that only revealed his wish to get "far away from the USSR." Ominously, Stalin read this letter carefully, as shown by his notes in the margins, and sent it on to the Politburo.[13]

Soon the entire Lenin Academy was overhauled, with nonscientists replacing President Vavilov and the vice-president. The task of the new leaders was to focus the institute on the practical solutions that Lysenko said he offered, rather than on the genetic research of Vavilov and his colleagues. To explain his alternative approach, Lysenko wrote a pamphlet on plant breeding that claimed that by altering the temperature and the amount of light a plant received, an early-ripening type could be transformed into a late-ripening one and vice versa, all in one generation. It was simple, at least to Lysenko and other believers in the inheritance of acquired characteristics: just select the early-ripening plants and discard all the others: Voilà! You have changed all their offspring forever.

At a 1935 meeting of "collective Farm Shock Workers," which Stalin attended, Lysenko explained the difference between bourgeois science, whose adherents only "observe and explain phenomena," and socialist science, which aims to "alter the plant and animal world in favor of building of a socialist society." Lysenko confessed to being "not an orator ... only a vernalizer," at which point Stalin responded, "Bravo, Comrade Lysenko, Bravo."[14] Government officials followed suit, with Yakovlev lauding the scientist who had "opened a new chapter in agricultural science."[15] This showed that with Stalin's endorsement, pseudoscience had become tacit state policy in the Soviet Union. Over the next nearly three decades, it would come to dominate Soviet science.

Michurinism: Lysenkoism Rebranded

Lysenko also recommended crossbreeding even the most carefully bred and developed plant strains, arguing that this would reinvigorate them. Vavilov knew that this would instead cause their offspring to lose the very characteristics the plants had been bred for, something that was obvious to anyone who accepted and understood genetics. In a speech at the October 1935 meeting of the Lenin Academy, Lysenko claimed the mantle of plant breeder Ivan Vladimirovich Michurin (1855–1935), who had died the previous July, as the authority for his views. "Our first task is to master [his] legacy," said Lysenko, "and demand ... a thorough knowledge" of his works.[16]

The life of Ivan Michurin, as historian David Joravsky wrote, "Began as a bleak variant of *The Cherry Orchard*."[17] Michurin's once-well-to-do family had fallen on hard times and had to sell fruit from their own orchard, but the venture failed. Michurin's mother died of tuberculosis, and his father, who is said to have sung a dance song rather than the expected lament at her funeral, had to be taken to a madhouse. Young Ivan was the only one of his seven siblings to survive this early period. He became a railroad worker, providing him with the opportunity to travel widely and visit many of Russia's most famous gardens. He grew so interested in plants that he quit his railroad job to start a tree nursery. Using crossbreeding, artificial selection of plants with the most desirable traits, and hybridization with wild plants, Michurin developed some 130 new varieties of fruit trees and berries. He wrote some hundred articles in Russian horticultural journals.

When Michurin's successes came to the attention of Lenin in 1922, he asked for a report on Michurin's work, which led to a visit by a high-ranking Soviet official. In 1923, Michurin's achievements were presented in Moscow at the First All-Union Agricultural Exhibition. Soviet journalists adopted Michurin as a twentieth-century folk hero like plant-breeder Luther Burbank and inventor Thomas Edison, who performed miracles sans diploma.

Like Burbank, Michurin denied that genetics explained heredity. He wrote that "heredity does not yield and in essence cannot conform to any patterns worked out by theoretical science and determined in advance." Those academics who claimed otherwise he denounced as the "caste priests of jabberology."[18] Late in life, however, Michurin came to understand Mendel's achievement. Perhaps plant breeders like Burbank and Michurin, untrained in and deeply skeptical of genes and genetics, had to believe that something was behind their breeding success—and inheritance of acquired characteristics seemed the only alternative. Moreover, given their trial-and-error methods, Lamarckism seemed to make a certain amount of sense. There was nothing wrong with what they did; indeed, both had many successes in breeding new plants. The results were of practical use but made no lasting contribution to science.

Isaak Prezent, who wrote about the relationship between Marxism and science, had become an ally of Lysenko, and with Michurin's death the two saw an opportunity. The term *Michurinism* had not yet come into wide use, but Prezent came up with the idea of using it as a catchall for Lysenko's many theories. The two carefully culled Michurin's writings so that they could present him as a staunch opponent of Mendel and other geneticists.[19] Lysenko wrapped himself in Michurin's mantle and tended hereafter to refer to his (Lysenko's) "theories" as Michurinism, a name and a method that would soon gain worldwide fame. Mao and the Chinese Communists would adopt it wholesale with even more lethal results.

In December 1935, Vyacheslav Molotov, Stalin's right-hand man and chairman of the Council of People's Commissars, spoke at a meeting of the leaders of the Lenin Academy. Vavilov had been removed from the presidency, but he still headed the Genetics Institute. Molotov read through the institute's research plan, where an item titled "On the Domestication of the Fox" caught his malevolent eye. A decade before, Vavilov had supported work with wild animals to explore the possibility of domesticating them, so he stood to respond to Molotov, who was not impressed. He began a tirade against the work of the institute under Vavilov, which he said had squandered money on a "collection of seeds that was of no use to anyone . . . and [left] to rot, anyway." In a later editorial, Molotov praised "the brilliant works of Academician T. D. Lysenko," and wondered ominously, "Isn't it strange that a number of scientists still have not found it essential to show Lysenko ... active support."[20]

Naming Names

At another meeting that December of "outstanding" farm workers, Stalin was again present. In his speech, Vavilov went out of his way to laud Lysenko, saying, "I must mention the brilliant work being carried out under Academician Lysenko's direction . . . a major world achievement in horticulture." Stalin did not hear this remark, for as soon as Vavilov began to speak, he rose ostentatiously and strode out of the hall. In his remarks, Lysenko complained several times about the scientists "who argue about the incorrectness of his [Lysenko's] methods." Finally, Yakovlev, head of the agriculture department of the Central Committee, urged, "And who exactly? Why not name them?" Lysenko replied,

> I could name names, although here it is not the names that are important, but rather the theoretical stand. Professor Karpechenko,

Professor Lepin, Professor Zhebrak [all closely linked to Vavilov], and generally the majority of geneticists disagree with our position. In the recently published work, Theoretical Principles of Wheat Breeding, Nikolai Ivanovich Vavilov, while agreeing with a number of theses advanced by us, also disagrees with our basic principle of rejecting defects during the selection process.[21]

The day after this meeting, Lysenko was awarded the Order of Lenin, the Soviet Union's highest honor. He would receive it seven times in all. Lysenko's rise in the Soviet hierarchy and obvious support from Stalin had made him increasingly "insolent, crude, and prickly."[22] Leading geneticists and plant specialists had tried to be polite, but frustrated by their fading relevance, demanded a showdown to decide who would prevail: Lysenko or Vavilov? The confrontation took place under the auspices of the Lenin Academy branch in Omsk in 1936. Just prior to the meeting, Vavilov and Lysenko had together inspected nearby fields in a "peaceable and friendly" fashion, but when the meeting commenced, Lysenko came out firing.[23] He said that Vavilov's claims of flax hybridization "did not occur, although Vavilov maintains that . . . he saw it himself with his own eyes." This was nothing less than an accusation of scientific misconduct. He continued, "Until the past few days, Academician N. I. Vavilov, Professor Karpechenko, and a number of others have nakedly and in an unprincipled manner spurned the fundamental pattern of development of hybrids that has been revealed to us."[24] As noted earlier, Karpechenko would be executed on July 28, 1941.

Lysenko continued his assault at the fourth plenary session of the Lenin Academy in December 1936. Vavilov had generously, if naively, believed that Lysenko was a successful practitioner whose problem was that he had never been taught genetics and whom others were exploiting. Vavilov thought that Lysenko's ignorance could be remedied and asked American geneticist H. J. Muller, now working with Vavilov, to prepare an exhibit explaining the science of genetics. Lysenko spent five minutes peering through the microscopes, then departed without saying a word.

Several speakers subjected Lysenko to the sternest criticism he had yet received, but again Vavilov was conciliatory and diplomatic.

Lysenko did not return the favor. Again, he named Vavilov as the chief opponent of his theories and repeated his claim that plants could be "trained" to change the traits they inherited. He cited the example of a single wheat plant that had been transformed into spring wheat over several generations by changing the temperature and moisture of its environment. This proved, Lysenko said, that immutable genes did not exist. But as historian Zhores Medvedev points out, "To discuss this experiment as a scientific one is absolutely impossible, since it was an experiment with a single plant, offspring of a single individual, a single seed. An experiment without replication is not a scientific experiment. The single seed could have been a hybrid, a mutant, or a contaminant. One casual seed does not represent a variety."[25]

Lysenko went on to claim that by dropping the concept of genes, his opponents "would easily come to the conclusion that winter plants at certain moments in their life, under certain conditions, can be transformed, can change their hereditary nature into a spring type, and vice versa—which we are quite successfully doing experimentally now." He continued, "By taking our point of view, N. I. Vavilov, too, will be able to refashion winter plants into spring ones. Moreover, any winter variety, in any quantity, can be made into a spring one." This was too much for Vavilov, who rose from his seat and asked of Lysenko, "You can refashion heredity?" Lysenko replied, "Yes, heredity." He went on, "Unfortunately, the concept of genetics, with its immutable genes in a long line of generations, with its non-acknowledgement of natural and artificial selection, nevertheless holds sway in the minds of many scientists."[26] The last was another lie, as geneticists did not deny natural selection but rather saw genes as the mechanism by which it operated. The meeting was a triumph for Lysenko, who arrogantly and sarcastically shrugged off dissent. Press accounts barely mentioned Vavilov's criticisms, while printing Lysenko's speech in full. Lysenkoism was in full bloom and Soviet agricultural science had become entirely an issue of ideology in which scientific evidence no longer counted.

Thanks to Vavilov's reputation, which remained untarnished outside the USSR, and to the strong impression he had made at the 1932 Sixth International Genetics Congress, the seventh had been scheduled for Moscow in 1937. At the meeting, Vavilov was to assume the presidency of the congress. The conference had been announced in scientific journals worldwide, the program had been printed, and some 1,700 scientists had said they would attend. Here was a chance for Vavilov to demonstrate that in every other advanced country genetics was viewed as an indispensable field of science. Vavilov's election as president would have enhanced his stature inside the Soviet Union and perhaps have continued to protect him from the terror that had already befallen so many other Soviet scientists. But on August 2, Stalin headed off these possibilities by telling Molotov to cancel the meeting, saying that "people in charge of preparing the congress are useless, they came up with a proposal before they prepared the business."[27]

4
Pseudoscience Defeats Science

"We shall go to the pyre, we shall burn, but we shall not retreat from our convictions."[1]

—Vavilov

Given his unconditional support from Soviet officials, up to and including Molotov and Stalin, one would assume that Lysenko must have produced many research successes and greatly enhanced Soviet agriculture. The truth was just the opposite.

False Claims and Failed Experiments

Lysenko's reportedly miraculous winter crop of peas in Azerbaijan in 1927 evidently was never repeated. His attempt to vernalize winter wheat failed in its first year, at which point, without explanation, he switched to promoting the vernalization of spring wheat. This too failed, but one would never have known it, as Lysenko praised his own methods in newspapers and at conferences. In 1932, Lysenko was supposed to have overseen the planting of 110,000 hectares (271,000 acres) of vernalized wheat seed, but only about 20,000 hectares (50,000 acres) were claimed to have been sown. And even that figure is suspect, as it came from extrapolations by Lysenko's assistants, based on questionnaires sent to only fifty-nine farmers whose combined acreage came to only 300 hectares (741 acres). Nevertheless, this was good enough to inspire Soviet plans to plant "thousands of hectares, then about tens of thousands, and finally about millions," of amber fields of vernalized wheat.[2] Lysenko claimed to have made rigorous tests of vernalization, but when scientists at the Lenin Academy tried to replicate his claim in 1936, they found that "on the average, both decreases and increases [in yield] from vernalization were observed, but vernalization produced almost no increment in the five-year average."[3] By 1937, the amount of farmland cultivated using Lysenko's methods had increased 163-fold, but actual food production declined and famine returned.[4] Vernalization was a net loss, as time spent on the laborious process could have been more productively used for traditional methods.

Another of Lysenko's strategies was to promise to develop new varieties of more resistant and higher-yield wheat. He pledged that in little more than two years his Odessa Institute would develop four new varieties of spring wheat, whereas breeding and testing new varieties by standard methods would normally take much longer. When that promise failed, Lysenko offered the prospect of creating a new variety in one year. Then, at a 1935 meeting in the Kremlin, he diverted attention again, now promising to come up with a new and improved variety of cotton in only two years. At a meeting of "labor heroes" a few months later, he repeated this promise and said, "If I do not obtain 40 kilograms of seeds of the new variety [of cotton] by springtime, then by the summer I would not be able to provide five tons, which would mean that I could not fulfill the promise given in the Kremlin to Comrade Stalin... and if I do not fulfill this task, what right will I have to call myself a scientist?"[5] This bold promise was also broken.

As these claims and subsequent failures were occurring in the mid-1930s, Lysenko switched his attention to another problem of Soviet agriculture: potato production. The potato is native to the Western Hemisphere, originating in the highlands of southern Peru, where the conquistadors encountered it and brought it (and many other crops) back with them. Today one can go to dinner in Peru and find many potato varieties on the menu, of all shapes, sizes, and colors. The potato quickly spread across the globe, becoming such an important foodstuff that by one estimate it alone made possible one-quarter of the growth in world population and urbanization between 1700 and 1900.[6] But when the potato is planted in a hot, dry climate, it is subject to diseases and harvests suffer. Scientists in the late nineteenth century discovered that the cause was a virus. Until then, potatoes had been vaguely said to suffer from a "degenerative" disease—a kind of enfeeblement. The hot, dry climate of the southern steppe of the Soviet Union was detrimental to potatoes, requiring that they be grown in the north and shipped to the south.

The traditional method of growing potatoes was to plant tubers, but they tended to be infected by any virus in the parental plant. But true seed does not have this defect, so that potatoes grown from seed are much heartier. Soviet scientists learned how to grow seed potatoes and certified those varieties that showed improvement. But as of 1931, only about 1 percent of the Russian potato crop was grown from certified seed. Potato specialists offered solutions, but in the cautious mode of the scientist, they pointed out the advantages and disadvantages of the proposed methods and called for field trials that would take years. The Lysenkoists responded by claiming that plant viruses do not exist, regarding them, in Joravsky's words, as "a metaphysical construct of bourgeois science, comparable to genes."[7]

Once more came Lysenko with a new solution to yet another agricultural problem. Without benefit of experiments, he claimed that a potato tuber planted near the beginning of autumn, when the soil is cooler, would improve in only one or two seasons. Again, Lysenko did not claim merely that individual plants would do better, but that, as with vernalized wheat, the "breed" itself would be improved. The Commissariat of Agriculture responded with a massive program of potato planting in the summers from 1936 through 1941, until the German surprise attack in Operation Barbarossa forced a halt. Lysenko evaluated the first year of the program by sending a questionnaire to 500 to 600 farms but published the responses of only the most favorable fifty.[8]

None of the potato specialists and virologists publicly questioned the practical value of summer planting, and a few professed to be enthusiastic Lysenkoists. Many half-heartedly endorsed Lysenko's method of late summer planting while ignoring the vexing subject of degenerative diseases. Others lauded Lysenko's practical success, while pointing to his lack of a theory to explain his results. As Joravsky writes, "Lysenko made short work of such critics. If their theoretical understanding was so great, why did they have no practical solution for degeneration? If his theory was so poor, how had he accomplished a cure?"[9] After the war, summer potato planting resumed among the farmers in the south but failed to increase yields. When the program received an unfavorable review in 1952, it was quietly dropped.

Gathering Clouds

The year 1937 marked the apex of Stalin's anti-Trotskyite terror. Nikolai Bukharin, the theorist of the party and close ally of Stalin, was arrested and shot as a spy and saboteur. To save his own skin, Nikolai Gorbunov, who had been Lenin's secretary and Vavilov's boss and supporter, now charged Vavilov with supporting fascist theories of genetics. But this came too late to save Gorbunov from the terror and he too was arrested and shot. I have already mentioned Nikolai Koltsov, one of the world's leading biologists, who was accused of supporting fascist eugenics and poisoned by the NKVD, the deadly successor to the OGPU. Vavilov's replacement as president of the Lenin Academy, Alexander Muralov, was arrested and shot in March 1938. The fate of Y. A. Yakovlev, appointed in 1929 as people's commissar for agriculture and transferred in 1934 to party headquarters as head of the agricultural department, is of special interest because we have Nikita Khrushchev's memoir to reveal the frighteningly arbitrary nature of Stalinist terror:

> After a meeting of the Central Committee Plenum in 1937, the attendees were dispersing for dinner when Stalin spotted Khrushchev and shouted, "Khrushchev, where are you going? ... come with me, we'll eat together." As we were leaving, Yakov Arkadevich Yakovlev, who had been hovering around nearby, followed Stalin to his apartment uninvited. The three of us ate dinner together. Stalin did most of the talking. Epstein-Yakovlev [Khrushchev maintained that Yakovlev's real name was Epstein] was in a very agitated state. You could see he was undergoing some sort of inner turmoil. He feared that he was about to be arrested. He wasn't mistaken in his forebodings. Shortly after Stalin's friendly chat with him over dinner, Yakovlev was arrested and eliminated. I'm telling this story to show how even someone as close to Stalin as Yakovlev—who was head of the Agricultural Section of the Central Committee and who had been

one of Stalin's most trusted supporters during the struggle against the opposition—could suddenly find his life hanging by a thread. The episode was typical of Stalin's treacherous character.[10]

Yakovlev was arrested on October 12, 1937, and his wife the next day. She was charged with being a French spy, and Yakovlev was told that she had denounced him. He was also charged with being a spy, but in his show trial in March 1938 was "transmogrif[ied] into a Rightist ... an odd appellation for the man who had been a chief operator in the collectivization field."[11] Yakovlev was executed on July 29, 1938.

Joravsky lists 105 biologists and agricultural specialists who were "repressed," most of them in 1937.[12] They were seen as direct threats to Lysenko or as standing in the way of his advancement. In the spring of 1938, Lysenko replaced Vavilov as president of the Lenin Academy. One of Lysenko's first acts was to propose cutting the scientific staff of the institute by half or more. Those who remained, he declared, "were to draw the deepest generalizations and conclusions from [collective and state-owned farms]."[13] In other words, they were to obey Party ideology rather than to try to exert scientific leadership.

On May 17, 1938, in speech to leaders of educational institutions, Stalin toasted:

> The progress of science, of that science whose devotees, while understanding the power and significance of the established scientific traditions and ably utilizing them in the interests of science, are nevertheless not willing to be slaves of these traditions; the science which has the courage and determination to smash the old traditions, standards and views when they become antiquated and begin to act as a fetter on progress, and which is able to create new traditions, new standards and new views. (Applause.)[14]

This was an implicit endorsement of Lysenko and his rejection of genetics—a *Pravda* editorial two days later made that clear—but Stalin himself did not make it explicit. Rather, for his own reasons he allowed the tension between the pro- and anti-Lysenkoites to fester, though no one could imagine that he favored the antis. After all, the year before, to clear the way for Lysenko to ascend to the post, two successive presidents of the Lenin Academy of Agricultural Sciences had been removed.

By December 1938, the NKVD's file on Vavilov filled five folders; now it opened a sixth, titled "Genetics." One of the first documents in this new file was labeled "Struggle Waged by Reactionary Scientists against Academician T. D. Lysenko." It claimed that Vavilov was "using every effort to discredit Lysenko as a scientist" and called for his arrest. A deputy of the notorious Lavrentiy Beria, the head of the NKVD and Stalin's accomplice in mass murder, had signed the document. But likely because Stalin was not yet prepared to go after the best-known Soviet scientist internationally, no arrest followed.[15]

Vavilov could not have failed to be aware of the machinations going on around him, but he soldiered on. Nearly five decades later, in 1987, the daughter of one of Vavilov's colleagues from long ago and a friend of his son Oleg looked back over the decades to recall a visit she made to the Vavilov home around this time: "The clouds were gathering," she wrote. "Whenever [Vavilov] took the train in Leningrad he wondered whether he would reach Moscow. He warned us children to watch our tongues."[16]

We Cannot Go on This Way

At a March 1939 conference of plant breeders and geneticists, Lysenko urged liberation from the Mendelism that "still sits within every one of us." The head of seed production announced that future work in plant breeding and seed production would be based on "Darwinism and Michurinism" and the Minister of Agriculture agreed.[17] The appeal to Darwin referred to his "soft inheritance of acquired characteristics," thus putting him within the scope of Lysenkoism, according to its followers.

Vavilov responded that Lysenko's supporters "were discarding the experience of world science without any basis . . . while world science follows the path laid down by Mendel, Johansson, and Morgan." (Wilhelm Johannsen (1857–1927) was a Danish geneticist known for coining the terms *phenotype* and *genotype*: the genetic makeup of an organism). Lysenko said that Vavilov's statements at the meeting were "so unclear, so scientifically vague, that one can only conclude from them that 'varieties fall from the sky.'" He even managed to associate Vavilov's views with fascist eugenics in Germany.[18]

At a meeting of scientific workers the same month at the All-Union Institute for Plant Breeding and Genetics (AIPB), Vavilov gave a speech that can stand as his scientific manifesto and his epitaph:

> There are two positions, that of the Odessa Institute and that of the AIPB. It should be noted that the AIPB position is also that of contemporary world science, and was not developed by fascists but by ordinary progressive toilers. ... And if we had here an audience of the most outstanding breeders, practical and theoretical, I am sure they would have voted with your obedient servant and not with the Odessa Institute. This is a complex matter. It is not to be solved by decree of even the Commissariat of Agriculture. *We shall go to the pyre, we shall burn, but we shall not retreat from our convictions.*[19] (Italics added.)

On April 17, 1939, the day after Lysenko's election to the Academy of Sciences, Vavilov published an article in *Socialist Agriculture* in which he said, "To those who propose the elimination of modern genetics we say: first offer a substitute of equivalent value. Let chromosomal theory be replaced by a better theory, not a theory that sets us back seventy years."[20] But Lysenkoism, even renamed Michurinism, had no underlying theory.

At yet another of the innumerable Soviet meetings, this one on May 25, 1939, the Lenin Academy, now under Lysenko, received a report from Vavilov on the work of the AIPB, of which he (Vavilov) was still director. On Lysenko's motion, the academy rejected the report and adopted a resolution declaring the work of the AIPB unsatisfactory. Then followed a long and demeaning Kafkaesque interrogation of Vavilov, as fully recorded by a stenographer.[21] In one exchange, Vavilov had admitted that "I am an overburdened man; not only do I work as the academic secretary, the deputy, but even as a financial administrative assistant. ... I should have explained this in greater detail. ... If Trofim Denisovich would only listen calmly instead of shuffling pages." In his concluding remarks, Lysenko said:

I agree with you Nikolai Ivanovich, it is somewhat difficult for you to carry on your work. We talked of this many times and I was sincerely sorry for you. But you see, you are being insubordinate to me.... I say now that some kind of measures must be taken. We cannot go on this way.... We shall have to depend on others, take another line, a line of administrative subordination.[22]

Soon Vavilov left on yet another collecting trip, ceding the stage to Lysenko, who wasted no time dissolving the entire scientific council of the AIPB, most of whom were Vavilov supporters. From that low point, matters descended. In October 1939, at the behest of the Central Committee, what would be the final confrontation of Vavilov and Lysenko took place. The arrangements for the meeting were left to the editorial board of *Under the Banner of Marxism*, a Soviet philosophical and socioeconomic journal. This ensured that the outcome would reflect politics, not science. It went on for a dreadful week. Especially considering Lysenko's ominous remark that "measures must be taken," Vavilov knew that he now had nothing to lose. "Lysenko's position not only runs counter to the group of Soviet geneticists, it runs counter to all of modern biology," he said. "In the guise of advanced science, we are advised to turn back essentially to scientifically obsolete views out of the first half or the middle of the 19th century.... What we are defending is the result of tremendous creative work of precise experiments by Soviet and foreign practice."[23]

Lysenko responded, "I do not recognize Mendelism. I do not consider formal genetics a science. And Vavilov... stand[s] in the way of objectively and correctly getting down to the essence of Mendelism, exposing the false and contrived nature of the... doctrine and ending the teaching of it in higher educational institutions as a positive science." Another speaker told geneticists that they "could bring great benefit... to the cause of socialism... if you renounce the rubbish and dross that have piled up in your science."[24] *Pravda* reported at length on the conference and *Under the Banner of Marxism* printed a complete transcript. The blatant denunciation of geneticists was regarded as an unofficial party directive and thus had to be followed.

A month after the conference, on November 30, 1939, Stalin summoned Vavilov. It would be the scientist's last meeting with the dictator. As Vavilov told a friend, as he entered the room Stalin was walking back and forth, head down, omnipresent pipe gripped in his hand. He neither looked up or responded to Vavilov's greeting. To break the silence, Vavilov began to report on the work of the Lenin Academy. Stalin continued to pace and after a few minutes went to his desk, sat down, and interrupted Vavilov mid-sentence to say, "Well, Citizen Vavilov, how long are you going to go on fooling with flowers and other nonsense? When will you start raising crop yields?" Vavilov began to explain the science that would be necessary to achieve that goal. Stalin listened for a few minutes and said simply, "You may go."[25]

Lysenko Towers Above

As historian Valery Soyfer points out, "In addition to being a theorist, experimentalist, teacher, and propagandist, Lysenko now began to take an increasing role in the Soviet government."[26] He had become so popular and so well recognized in newspapers and magazines that it was thought only natural for him to assume a broader responsibility beyond science. In 1936, Lysenko attended the Eighth Congress of Soviets and joined the editorial commission that produced the final text of the "Stalin Constitution." The next year he became a member of the USSR Supreme Soviet and the year after that was made vice chairman of one of its two chambers, the Council of the Union. This gave Lysenko an ostensible rank higher than that of Stalin himself, who was a member only of the executive body, which lacked legislative power.

When the Supreme Soviet met at the Great Kremlin Palace, officials were placed on one of three platform levels by rank. "On the lowest were seated the party and government leaders, including Stalin; the rostrum for speakers was positioned a little higher; and at the very summit were the chairman and vice chairman of the two legislative chambers."[27] For a number of years, Lysenko literally looked down on Stalin. In his role as vice chairman of the council of the union, Lysenko presided over sessions that had nothing to do with science or agriculture. For example, he was present when in early August 1940, the Supreme Soviet approved the annexation of parts of Finland,

Estonia, Latvia, Lithuania, Poland, and Ukraine, as outlined in a secret protocol of the Molotov-Ribbentrop Pact between the USSR and Germany signed on August 23, 1939, one week before Germany invaded Poland. As Soyfer writes, "In the steady stream of front-page photographs of the Supreme Soviet, every citizen of the country could see Lysenko seated, standing, applauding, but always towering above Stalin, Molotov, Beria, Voroshilov, Khrushchev, and Vishinski. There was no higher place of honor."[28] Could there be any clearer evidence that, with Stalin's approval, Lysenkoism had become de facto state policy in the USSR?

The Death of Vavilov

With the world's attention focused on the war, in August 1940 the NKVD arrested Vavilov while he was in Ukraine on a field trip to study the territory that the USSR had just acquired through the Molotov-Ribbentrop secret protocol. He was taken to a cell in Moscow's notorious Lubyanka prison, where he was tortured and interrogated nearly 400 times, and then given only one day before his trial to study the charges against him. Vavilov confessed to having been a member of a "rightist" organization in the Commissariat of Agriculture and was forced to name other scientists as his co-plotters. He identified a dozen leading Soviet agriculturalists, all of whom were already under arrest or had been shot. His five-minute show trial supposedly featured testimony from Alexander Muralov, who had replaced Vavilov as president of the Lenin Academy. Vavilov was able to get his death sentence commuted to twenty years of "deprivation of freedom" and was sent to the prison hospital in Saratov, where two decades before he had begun his career as a young professor. But it was too late, as he was already suffering dystrophic diarrhea and dying of starvation. On January 23, 1943, Nikolai Vavilov died, and his body was dumped in a common grave.[29]

In 1955, after the death of Stalin, the Central Committee of the USSR began to review cases that had led to the death penalty. In August, the committee rescinded Vavilov's conviction after the prosecutor pointed out that, for one thing, Muralov had been shot three

Figure 4.1 Vavilov's prison mugshot
Wikimedia Commons

years before his alleged testimony. Vavilov's wife, Yelena Barulina, was the first to be notified of her husband's exoneration and rehabilitation. She began work on his manuscripts, including his World Resources of Cereals and Flax and Their Application for Selection, which was published in 1957, one month before her death. Publication of his *Five Continents: Tales of Exploration in Search of New Plants* followed in 1962. Today, his name lives on in the history of biology and genetics and at the N. I. Vavilov All-Russian Institute of Plant Genetic Resources in Leningrad and at the N. I. Vavilov Institute of General Genetics in Moscow. (See Figure 4.1.)

5
Rise and Fall

"Stormy applause. Ovation. All rise."[1]

Audience response to Lysenko's 1948 announcement that the USSR had banned the study and practice of genetics

With Vavilov in his grave, the way appeared open for Lysenko to take full command of Soviet biology and agronomy. Then came World War II, an existential struggle for the USSR to which everything else had to take a backseat.

Sunflowers into Strangleweed

As vast areas of the western Soviet Union were lost to the Wehrmacht juggernaut, grain production in the Soviet Union shifted east, and millet, a more resilient crop with a shorter growing season yet almost as much nutritional value, replaced wheat on many farms. Although total grain production in the USSR fell during the war, famine was largely avoided. This was due to several factors: an increase in the number of machine-tractors, the patriotic devotion aroused by the attack on the nation, the switch to millet, the increased role of women and children in farming, and the reclamation of western lands as the Germans retreated.

As the Wehrmacht drove eastward in the first year of the war, seemingly invincible, Soviet scientific institutes and factories were evacuated to locations beyond the barrier of the Urals. Lysenko moved to the Siberian Grain Institute in Omsk in southwestern Siberia, 2,700 kilometers (1,670 miles) east of Moscow. The productivity of Soviet agriculture had become a life-or-death issue for the USSR, and there was no time for the interminable and fruitless agricultural debates that had marked the 1930s. Now Lysenko and other agronomists had to deliver practical results, and fast. Although his wartime record was mixed, Lysenko did have scattered positive results. Soviet citizens were encouraged to grow "victory gardens," most of which were planted in potatoes. Lysenko made the practical suggestion that the eyes of the potatoes be planted and the rest eaten, a recommendation that was printed in the newspapers and saved many lives. When Lysenko encountered the frigid fields of Siberia, he realized that much of the wheat crop could be lost to frost and snow and recommended that it be harvested before ripening, saving much grain.

On the other side of the ledger, Lysenko recommended planting sugar beets in arid Uzbekistan, where they withered and died. He resurrected an older idea by proposing that local wheat varieties be planted directly on the stubble remaining after the cutting of the spring wheat, which supposedly would protect the newly planted seeds.[2] The director of the Omsk institute objected, prompting Lysenko to appeal to the USSR Commissariat of Agriculture, who backed him and ordered the technique applied throughout Siberia. It never worked, and in 1956 a Party journal admitted as much, writing that "in the Omsk region alone, tens of thousands of hectares sown on stubble never yielded even so much as the seed that was planted."[3] The article failed to note that the Party Commissariat had approved the method.

After the war, Lysenko renewed his attack on Darwinian biology. Vavilov and most of his other enemies were dead or disgraced and no longer in a position to impede him. Lysenko now took on the field of cytology: the fundamental branch of biology that deals with the structure and function of cells and whose findings strongly support genetics. Lysenko spoke approvingly of the work of an elderly Bolshevik doctor named Olga Lepeshinskaia (1871–1963). For years, she had denounced cytology and claimed that she could create cells starting from egg yolks or from other noncellular organic materials. Lepeshinskaia believed in *spontaneous generation*: that living creatures can develop from nonliving matter. Lysenko and Lepeshinskaia both rejected the essential process of *meiosis*, in which a cell divides into four, each part having half the number of chromosomes of the cell from which it came. For this pseudoscience she was celebrated, winning the

Stalin Prize in 1950. Next Lysenko turned his ire on the theory of natural selection, saying that he had discovered that its fundamental concept is false: organisms of the same species do not compete. Just the opposite: weaker tree seedlings, for example, surrender themselves so that the stronger can survive.

Lysenko and his followers would go on to ever more extreme Lamarckist claims. Historian Soyfer reports that in a 1955 lecture at Moscow State University, he heard Lysenko "describe how the lazy cuckoo placed its eggs in the nest of a warbler, and the warbler is then compelled by the 'law of life of a biological species' to pay for letting the cuckoo take advantage of it by feeding on caterpillars, and as a result of the change in diet, hatches cuckoos instead of warblers."[4] Soyfer notes that instead of the laughter that this claim should have provoked, "the majority of students seemed to believe it." Agronomist and biologist Zhores Medvedev wrote that "in nearly every issue of [Lysenko's journal *Agrobiologiya*], articles reported transformations of wheat into rye and vice versa, barley into oats, peaches into vetch, vetch into lentils, cabbage into swedes (rutabaga), firs into pines, hazelnuts into hornbeams, alders into birches, sunflowers into strangleweed."[5]

The 1948 Lenin Academy Conference

During the 1930s, Stalin and the Communist Party had chosen to isolate biological science in the USSR from that prevailing in the rest of the world in favor of its native "Michurinist" version. Nothing could have made this policy more vivid than two meetings held in the summer of 1948. In early July, the Eighth International Congress of Genetics convened in Stockholm for its first postwar meeting.[6] As Bengt Bengtsson and Anna Tunlid sum up, this was surely one of the most important conferences in the history of the organization, marking as it did not only "the commencement of normal scientific exchanges after the defeat of Nazi Germany and its allies in the Second World War" but also "the fight against the Stalinist version of hereditary science, Lysenkoism. This hotchpotch of ideas in which the inheritance of acquired characteristics played an important part had been lying low since the beginning of the war but was now again raising its head."[7]

American geneticist and Nobel laureate H. J. Muller (1890–1967) gave the presidential address and, no doubt to the surprise of the audience, devoted only a few words to the recent defeat of Nazism. Instead, he spent most of his speech on a new and growing struggle: the need to defeat Lysenkoism, which he had seen up close during his years working with Vavilov at the Lenin Academy.

The other meeting took place between July 31 and August 7, 1948, when the Lenin Academy convened in Moscow. Lysenko, now the Academy's president, organized the conference and Stalin approved it. In the immediate postwar years, Lysenko had begun to come under increasing criticism from a new generation of Soviet biologists and agricultural experts. In November 1947, he published an article in a popular magazine in which he again denied that individuals of the same species compete. The dean of the Biological Faculty of Moscow State University and other prominent biologists took issue with this anti-Darwinist claim. Lysenko must have felt the heat, for he wrote to Stalin asking for his protection from the geneticists who were attacking him, saying that "Mendelism-Morganism, Weissmanist Neo-Darwinism . . . are not developed in Western capitalist countries for the purposes of agriculture, but rather serve reactionary purposes of eugenics, racism, etc. There is no relationship between agricultural practices and the theory of bourgeois genetics."[8] The first reference above was to American Thomas Hunt Morgan, who was the first to use fruit flies to study genetics and who in 1933 won the Nobel Prize in Physiology and Medicine for showing that chromosomes carry genes. As for Weismann, in 1890, before the rediscovery of Mendel's experiments, he had found that inheritance occurs only by means of *germ cells*: egg and sperm cells, and not via other cells of the body (somatic cells), as Lepeshinskaia had claimed. This meant that acquired characteristics cannot be inherited, helping lead the way to modern genetics.

Stalin wrote to assure Lysenko of his continuing support, saying, "As for theoretical concepts in biology, I think that Michurin's concept is the sole concept that is scientific. Weissmanists and their supporters, who deny inheritance of acquired properties, do not deserve that we go on about them for long. The future belongs to Michurin."[9] This encouraged Lysenko to seek even stronger support, promising Stalin that he could increase wheat production five- to tenfold using "branched wheat," a variety that other studies had already shown to be no better than common spring and winter wheat. Stalin responded by expressing his trust in Lysenko's "agronomic genius" and allowing him to declare as an official position that genetics was a "bourgeois perversion."[10] In July 1948, the Soviet Politburo unanimously banned genetics as a scientific discipline.

Lysenko drafted the statement that he intended to give at the 1948 Lenin Academy conference and sent it to Stalin. The dictator deleted whole passages and edited the remainder word for word, making it as much his own as Lysenko's.[11] The speech, titled "The Situation in Biological Science," took up the entire first day of the meeting. It disallowed genetic research in the USSR and prohibited even discussion of the topic. *Pravda* published the address on August 4, while the meeting was still underway. Most of the remaining speakers spoke with enthusiasm about Lysenko's contributions and his practical successes. In what Stephen Jay Gould wrote "may well be the most chilling passage in all the literature of twentieth-century science,"[12] Lysenko closed the conference by proclaiming:

> Comrades, before I pass to my concluding remarks, I consider it my duty to make the following statement. The question is asked in one of the notes handed to me, 'What is the attitude of the Central Committee of the party to my report?' I answer: The Central Committee of the Party examined my report and approved it." In the record of the meeting, the stenographer added, "*Stormy applause. Ovation. All rise.*"[13] (See Figure 5.1)

As Borinskaya et al. sum up, "Lysenko, as the organizer of the August 1948 session, is often called the diabolical genius of Soviet biology. However, he was only Stalin's minion. If Stalin had not directed it, the session would have merely been one of the scandalous campaigns of that time, affecting only the VASKhNIL [Lenin Academy]. However, Stalin's involvement gave it global significance."[14]

Immediately following the conference, the Minister of Higher Education ordered the firing of anyone who "actively fought against Michurinists and Michurinist doctrine and failed to educate the Soviet youth in a spirit of progressive Michurinist biology."[15] Thousands were

Figure 5.1 Lysenko addresses the August 1948 conference of the Lenin All-Union Academy of Agricultural Sciences.
Borinskaya et al.[a]
[a]Borinskaya, "Lysenkoism against Genetics."

dismissed, and Isaak Prezent was appointed head of the Department of Darwinism at Leningrad State University, despite lacking any credentials in biology. Leaders of higher education institutions were told to bring about "a fundamental restructuring of educational and research activities to equip students and researchers with knowledge of the ground-breaking, progressive Michurinist doctrine and to vigorously root out the reactionary idealistic Weismannist (Mendelist-Morganist) branch."[16] Textbooks on biology that failed to endorse Lysenko's views were destroyed, as were mutant strains and genetic samples from laboratories. Here we see state science denial moving from the halls of the Kremlin down to the biological institutes and universities, and eventually into the classrooms, laboratories, and textbooks.

In the biological sciences, the only way to avoid the terror, and perhaps not even then, was to swear allegiance to Lysenkoism. N. A. Maksimov, director of the Institute of Plant Physiology, who had given Lysenko's claims a cool review in 1929, was called before the Academy of Sciences after the 1948 meeting, where he professed "his limitless devotion to Lysenkoism."[17] He apologized for the scientists at his institute who had publicly disapproved of Lysenkoism and said that henceforth the institute would be purely loyal to the Lysenkoite school. Maksimov's contrition allowed him to escape the terror and die a natural death in 1952. One person who had criticized Lysenko before the meeting recalled in a 1987 speech the mood as the terror had spread throughout the biological sciences decades earlier: "You were not afraid for yourself, but feared that, failing to bear torture during interrogations, you would incriminate and ruin your friends and colleagues."[18]

Song of the Forests

In October 1948, two months after the close of the conference, Soviet newspapers proclaimed the "The Great Stalin Plan for the Transformation of Nature." (See Figure 5.2) One of its several aims was to plant trees in a huge network of "shelterbelts" across the flat grassland terrain from Ukraine to Kazakhstan. These would provide windbreaks that would slow the drying winds from Central Asia and ward off the drought they caused. Something similar had been tried in the United States following the Dust Bowl years of the early 1930s. One hundred and twenty million hectares of trees were to be planted in the USSR within seven years, an area equal to that of Britain, France, Italy, Belgium, and the Netherlands combined. As a portent of its likely failure, the project was put under the auspices of a new organization headed by a Lysenko protégé. At the 1948 conference, this disciple had declared, "There is no nonparty science. Michurinist biology is a principled, Party science and will not tolerate compromise [with bourgeois science.]"[19]

Lysenko had convinced himself that the easiest and cheapest method of improving the success of oak trees was to plant them in dense clusters. The plants, he said, knew how to improve themselves. They did not enfeeble each other by competing, but rather cooperated in the need for moisture, sunlight, and nutrients. The weaker oak seedings sacrificed themselves so that the stronger might survive. Competition took place only between species, not among them. So instead of planting acorns far enough apart so that the saplings would not interfere with each other, the traditional method, several acorns were to be put into holes placed close together. As Joravsky writes, "In one huge rush cold dry Russia was to be made a land of mild moist

Figure 5.2 Stalin pondering his reforestation plan. Translated, the text reads, "We'll conquer drought, too."
Wikimedia Commons

weather," as the millions of trees would create their own weather.[20] The trees would manage themselves!

In 1949, before the project had gotten underway, Dmitri Shostakovich composed an oratorio titled the *Song of the Forests* to celebrate the reforestation of the steppes. That same year, before the Great Plan had barely gotten underway, Lysenko received his third Stalin Prize of 200,000 rubles. This was not the first time that Lysenko had been celebrated before his claims had an opportunity to be tested; indeed, that was the rule. The following overblown encomium accompanied his 1949 prize:

> When we say "Academician Lysenko" ... before our eyes a majestic panorama of the near future of our socialist agriculture opens up. We see vast fields of spreading wheat, vineyards in the central provinces, flowering fields in the polar regions, citrus plantations in the Ukraine ... and in the torrid steppes ... and many other areas, we see green expanses of broad-leafed oaks, as well as birch. ... We see winter

Figure 5.3 Sculpture of Stalin and Lysenko, erected in Stavropol in 1952 and demolished in 1961 after Stalin's fall from grace. Stalin holds a branching head of wheat.
Courtesy of the Archive of Administration of Stavropol

wheat that has found its new motherland in the steppes of Siberia, [and] highly productive herds on collective and state farms.[21]

Eminent Soviet scientists professed to support the tree planting, but the climate and the trees did not cooperate, and by 1951–1952, the Great Plan was on the way out. By 1956, only 5 percent of the trees planted during the project were viable and these only because the planters had ignored Lysenko's cluster method.

At Khrushchev's Side

By the early 1950s, with Stalin's now explicit support, as exemplified in the statue shown in Figure 5.3, Lysenko had banned his opponents and taken complete control of Soviet biology and agriculture. But on March 5, 1953, Stalin died of a heart attack. Nikita Khrushchev succeeded him as first secretary of the Communist Party and held the post until 1964. On February 14, 1956, Khrushchev gave the opening remarks at the 20th Party Congress. No doubt attendees expected a hagiography of Stalin, but instead his name was merely one on a list of those who had died since the last Congress. Several days later, Khrushchev gave a four-hour oration, known implausibly as the "Secret Speech," in which he demolished Stalin's reputation. As Khrushchev recalled in his memoirs:

> The congress listened to me in silence. As the saying goes, you could have heard a pin drop. It was all so sudden and unexpected. Of course it must be understood that the delegates were thunderstruck by this account of the atrocities that had been committed against worthy people, against Old Bolsheviks and Young Communists. How many honest people had perished, people who had been promoted to work in various sectors![22]

In the years just prior to Stalin's death, Lysenko had begun to come under criticism, in part for his absurd Lamarckist claims. Then in 1953, with Stalin gone, the *Botanical Journal* called for reopening the subject of genetics for discussion. Khrushchev joined in the disparagement, charging Lysenko with falsifying experiments and attempting to make himself the dictator of Soviet science. In April 1956, newspapers tersely reported that Lysenko had resigned as president of the Lenin Academy. However, after Khrushchev visited Lysenko's experimental farm in 1954, he changed his mind. In a March 1957 speech, Khrushchev charged the leaders of the Academy of Agricultural Science with having "folded their hands like saints and refrained from intervening in this dispute," when they should have supported Lysenko's techniques. "There are scientists who still dispute Lysenko," he said. "If I were asked which scientist I vote for, I would say without hesitation: for Lysenko. I know he would not let us down, because he doesn't put his hand to bad things. I think few scientists understand the soil as does Comrade Lysenko."[23] In August 1961, Lysenko was reinstated as president of the Lenin Academy. A year later, the chorus of criticism had become so loud that Khrushchev had to agree to his dismissal, though Lysenko was allowed to choose his successor.

Lysenko suffered a damaging wound in 1962 when three prominent Soviet scientists publicly accused him of engaging in pseudoscience and using his political influence to silence and eliminate those scientists who opposed him. With Khrushchev's dismissal as first secretary in 1964, Lysenko no longer had a protector, and the president of the Academy of Sciences officially declared him no longer immune to criticism. By this time, Lysenko was director of the Institute of Genetics at the Academy of Sciences, but he soon lost the position, and the institute itself was dissolved. Lysenko was exiled to an experimental farm in the Lenin Hills outside of Moscow. In 1965, an official commission visited the farm to examine Lysenko's claims. Its report was devastating, and when it was made public, Lysenko's reputation was shattered beyond repair.[24]

Lysenko died in 1976, earning only a minor mention in *Izvestia*. Famed Soviet physicist and Nobel laureate Andrei Sakharov may have written the most fitting epitaph for Lysenko twelve years earlier, in 1964, when the General Assembly of the Academy of Sciences was considering the election to full membership of another biologist, Nikolai Nuzhdin, who had served under Vavilov but renounced him to embrace Lysenkoism:

> Together with Academician Lysenko, he [Nuzhdin] is responsible for the shameful backwardness of Soviet biology and of genetics in particular, for the dissemination of pseudo-scientific views, for adventurism, for the degradation of learning, and for the defamation, firing, arrest, even death, of many genuine scientists.[25]

Graham Visits Lysenko

One of the few Western academics to have met Lysenko in person is science historian Loren Graham (b. 1933), who visited Russia on many occasions to interview scientists.[26] In 1961, while Graham was a graduate student at Moscow University and Lysenko resided at the nearby farm, Graham asked to meet with him. Lysenko never responded. Graham's next attempt came in 1971, when despite Lysenko's fall from grace, he was still a member of the Soviet Academy of Sciences, drawing a good salary and substantial benefits. Graham's clever strategy for gaining an interview was not to ask a potential subject directly to meet with him, but rather to leave a draft copy of an article or book chapter he had written in the scientist's office, saying that it would soon be published in the United States and offering to meet with the scientist to correct any errors. This seldom failed, but the suspicious Lysenko proved the exception. Several weeks later, Graham happened to be having lunch at the faculty club of the Academy of Sciences. As he tells it, "Sitting at the back of the room was a gaunt and homely man. I immediately recognized Lysenko."[27] Graham moved to sit beside Lysenko and after a while introduced himself as an American who had written about him several times. Lysenko said he had read Graham's books but that they contained serious mistakes about him. Asked what those were, Lysenko responded, "You accuse me of being responsible for the deaths of many Russian biologists, such as the well-known geneticist Nikolai Vavilov. I am not responsible for what either the Party or the secret police did in biology." Graham cited the 1935 meeting when Lysenko had named Vavilov among the traitorous "wreckers." "You criticized him in the presence of Stalin," Graham responded, "and that criticism had fateful consequences. He died in the custody of the secret police." Lysenko got to his feet and left the room without responding.[28]

Bloodlands

What was the cost of the science denial practiced by Stalin and Lysenko in lives ended, careers destroyed, and education stultified? As to the death toll, we can only make an educated guess. A recent estimate comes from Timothy Snyder's chilling 2012 book, *Bloodlands*.[29] He reports that in 1937 a Soviet government census found eight million fewer people than demographic trends had projected, most of them victims of famine in Ukraine, Soviet Kazakhstan, and Soviet Russia, and children never born. Stalin hid these results and had the demographers executed. In private conversations in 1933, Soviet officials put the number who died from starvation in the three regions at 5.5 million.[30]

The recorded number of excess deaths from famine in Ukraine was 2.5 million, which must be low since as the Holodomor proceeded, the system for reporting broke down, and many deaths went unrecorded. Another study done by Ukraine after its independence estimated that 3.9 million died from famine. Snyder makes a good case that "no fewer than 3.3 million Soviet citizens died in Soviet Ukraine of starvation and hunger-related diseases; and about the same number of Ukrainians (by nationality) died in the Soviet Union as a whole."[31] This would bring the total close to seven million. Of course, even without Stalin and Lysenko, there would well have been some level of starvation and hunger-related diseases; famines were a frequent occurrence in these lands. These different estimates make it impossible to know how many of the deaths were due to their policies, but surely it was in the millions.

Then there are the unquantifiable costs of Lysenkoism. In the nation that prior to the Revolution had been a world leader in science, the post-Revolution rejection of genetics and Darwinian evolution essentially destroyed the field of biology. In other countries, beginning with the modern synthesis that married genetics and Darwinian evolution, extending through the discovery of DNA by Watson and Crick in 1953, and on to the rise of molecular biology, while Lysenkoism crippled biology in the USSR, elsewhere it underwent a revolution. But Soviet biologists were prohibited from even discussing these advances, much less contributing to them. An entire generation missed out on some of the most important developments in the history of the biological sciences.

By causing farmers to abandon traditional crops and methods that would have produced much better results, Lysenko made the famines worse. In schools and universities, the teaching of modern biology was forbidden, so the brightest students chose other fields. Those who would become doctors and nurses were forced to learn Lysenkoist pseudoscience instead of modern biology. Thousands of ambitious and ill-educated people, gullible or venal enough to accept Lysenkoism, came to hold leading positions in science and in universities. Even in the 1990s, Soyfer wrote, "Lysenkoism remains deeply entrenched in Russian science, despite the end of the USSR.

Many of those who were raised in the Lysenko tradition still occupy key positions in Russia's science administration. Lysenkoism left a baneful legacy in the form of these followers, who continued to teach and work in many universities and institutes."[32] One can only imagine what discoveries Soviet biologists, freed from Lysenkoism, might have made, some potentially lifesaving, to the benefit not just of their own society but to humanity.

As noted in the Introduction, science denial has sometimes become state policy without an alternative theory to take its place, as evidenced by today's global warming denial. Soviet Lysenkoism is the clearest example of the other type, in which pseudoscience replaces mainstream science. In the USSR, a particular set of circumstances came together to give Lysenkoism a head start and to keep it going for three decades. These were: (1) Stalin accepted the inheritance of acquired characteristics throughout his life. The hypothesis appealed to the Marxist view that not only crops, but Soviet man, could be permanently improved; (2) as shown particularly by Stalin's editing of Lysenko's 1948 speech, the dictator was willing to substitute his own judgement for that of scientists and to persecute, even execute those who disagreed; and (3) Lysenko's claims were rarely tested, and even then half-heartedly, giving his pseudoscience an apparent scientific respectability.

The Soviet Union was the model for other states that wound up behind the Iron Curtain, many of whom also adopted Lysenkoism, or Michurinism, as they preferred to call it. Far to the east, four years before Stalin's death, the Red Star rose over the new People's Republic of China. Its leader, Mao Zedong, would pursue his version of Soviet Michurinism to an even-greater holocaust.

6

Big Brother, Little Brother

"The Soviet Union's today is China's tomorrow."[1]
—Mao Zedong

As documented by Walter Mallory in his 1928 book *The Land of Famine*, throughout its history China has suffered from famine. According to Chinese scholars, from 108 BCE to 1911 CE, a staggering 1,828 famines took place, with some provinces facing near-annual occurrences.[2] When Mallory wrote his book, the Chinese were still dealing with the effects of the devastating drought of 1920–1921, which caused half a million to starve. Unfortunately, lethal famines continued, including one in Sichuan in 1936–1937 that resulted in the loss of five million lives. The deadliest of all was the Great Chinese Famine of 1959–1961, which took place during Mao's Great Leap Forward and caused tens of millions to die. Unlike previous Chinese famines, the Great Chinese Famine did not have a natural cause but resulted directly from the denial of mainstream science and the adoption of Lysenkoist and Maoist pseudoscience.

Mao and the Victory of Communism

In 1919, Mao Zedong (1893–1976), then a librarian at Peking University, helped found the Chinese Communist Party (CCP). Four years later, the Chinese Communists joined forces with Sun Yat-Sen's Nationalist Party (Kuomintang), and Mao became head of its propaganda department. After Sun's death in 1926, Chiang Kai-shek assumed leadership of the Nationalists and in one of his first acts expelled the Communists. Then began a two-decade struggle between the CCP and the Kuomintang, which continued even while both fought the Japanese occupiers during WWII. As Mao remembered it in 1962, Stalin at first gave only lukewarm support to the CCP:

> In 1945, Stalin wanted to prevent China from making revolution, saying that we should not have a civil war and should cooperate with Chiang Kai-shek, otherwise the Chinese nation would perish. But we did not do what he said. The revolution was victorious. After the victory of the revolution he [Stalin] next suspected China of being a Yugoslavia, and that I would become a second Tito [who refused to bend to Stalin.][3]

The Red Army won the struggle in 1949 when it captured the Nationalist capitol of Nanking, forcing Chiang Kai-shek and the Kuomintang to relocate to the island of Formosa, now known as Taiwan. In December of that year, Mao traveled to Moscow for a two-month negotiation to secure Stalin's support for the new nation, which led to a mutual assistance treaty and economic aid. On February 14, 1950, China and the Soviet Union signed a thirty-year Treaty of Peace, Security, and Friendship, as mythologized in Figure 6.1. China's subsequent entry into the Korean War on the side of North Korea convinced Stalin that Mao and his comrades were genuine international Communists after all.[4] The Soviets then assisted China in drawing up its first five-year plan, which went into effect in 1953. The People's Republic of China (PRC) would follow in the footsteps of its "elder brother," enabling the new nation to advance more rapidly toward genuine Communism. Unfortunately, one of the lessons learned from big brother was the rejection of genetics and the adoption of Lysenkoism as the basis for Chinese agricultural policy.

Michurinism Comes to China

Before the pact was even signed, in the late 1940s the Soviets launched a program to introduce the Chinese to "Michurinist" biology, which was nothing more than Lysenkoist pseudoscience under a less-controversial name. Two Soviet propaganda agencies, the Sino-Soviet

Figure 6.1 Treaty of Peace, Security, and Friendship
C. 1950
Chinese Posters Foundation 2019

Friendship Association and the All-Union Society for Cultural Relations with Foreign Countries, were primarily responsible for this knowledge transfer.[5] As we have seen, by whatever name Michurinism entailed a complete rejection of classical genetics. According to Michurinist doctrine, inheritance is a characteristic of the entire organism, not of discrete hereditary factors such as genes. In fact, in the Michurinist view genes do not exist and acquired traits are heritable.[6] The two Soviet agencies were eminently successful, propagating Michurinist pseudoscience everywhere behind the Iron Curtain and, wherever it could, beyond it. The exportation of Michurinism to China was part of the Soviet Union's larger agenda of creating a Communist state in East Asia to act as a buffer between itself and American power in the region.

As historian of China Lawrence Schneider has pointed out, Michurinist biology could have remained a small adjunct to Sino-Soviet friendship, but instead the Soviets chose to make it so prominent that to the Chinese, Michurinism became the very emblem of Soviet science, the USSR, and its people.[7] The sharing of Michurinism with the Chinese demonstrated the willingness of the Soviets to help China, whose success would provide a second example to prove that Marxist science was superior to that of the West. In the 1950s, China borrowed a large amount of money from the Soviets, and both parties believed that the agricultural surpluses that were sure to result from following Michurinism would help pay back the loans.[8]

In the late 1940s, only some 200 Chinese scholars had doctorates in biology, most of them from U.S. universities.[9] Approximately sixty had been educated in agricultural biology at Cornell. Some Chinese biologists had even completed their theses under the guidance of Nobel laureate and geneticist Thomas Morgan, while others worked with Theodosius Dobzhansky, who had forsaken the USSR for America and become a leading geneticist. During the conflict with Japan, these Chinese biologists and most academics relocated to the far west of China. Afterward, they returned to help rejuvenate Chinese universities and their careers. Most Chinese Communists, in contrast, spent the war years in the northeast of the country. Few were trained scientists, leaving no one to contest Michurinist biology when after the war it was introduced in northeast China, formerly Manchuria. Michurinism burgeoned in 1948 after the Sino-Soviet Friendship Association branch in Dalian published a biography of the old Soviet breeder. From there, Michurinism expanded into the largest mass organization in China, generating a torrent of pro-Soviet propaganda for the next eight years. China's Ministry of Agriculture translated, replicated, and distributed Soviet materials on Michurinist biology. Spearheading the effort in the early 1950s was Luo Tianyu, a CCP cadre who, like the Soviet Lysenkoists, believed that science and scientists must yield to political ideology and practical outcomes. His education consisted only of two years at a Chinese agricultural college. With grand fanfare, Luo and his associates founded the Michurin Study Society, which published a journal composed predominantly of articles translated from Russian, as China lacked a history of Michurinism and its own literature on the subject. One translated article was Lysenko's notorious 1948 speech at the Lenin Academy meeting, wherein he declared the ban on classical genetics in the Soviet Union.

Some of the most prominent and devoted Soviet Michurinists came to China to deliver lectures and convert leading Chinese geneticists, but they had little success among this largely American-educated group. All of this transpired without an official CCP policy on "Mendelist-Morganist" biology.

In 1950, a group of Michurinists gained control of the august Science Society of China, much as the Lysenkoists had taken over the Academy of Sciences in the USSR. At the annual conference of the Beijing Biological Studies Society in July 1950, Luo was given a full day on the agenda, which he devoted entirely to Michurinism. The translated Soviet literature charged that classical geneticists were not merely wrong but had supported racism, fascism, eugenics, and during the war, even Hitler himself. These absurd claims went too far and soon the CCP was receiving complaints about Luo from the Chinese Academy of Sciences and others. These reached the ears of Mao and Liu Shaoqi, the vice chairman of the CCP, who removed Luo from his position. With the advantage of hindsight, we can see that this was the first point of divergence for China from the path of the Soviet Union, where criticisms of Lysenko only succeeded in causing Stalin to intensify the defense of his protégé.

In 1952, the CCP consulted widely with scientists and educators as to how to resolve the split between the Michurinists and classical geneticists. One result was the humiliation of Luo in the *People's Daily*, the party organ, for having created divisiveness between Party members and non-Party intellectuals, instead of converting the latter. Luo's failing was blamed on his poor understanding of science, which it was said had made him unable to present a convincing case against Mendelism-Morganism.

The CCP's solution to ending the dilemma was simply to declare one side the victor, announcing that "in the new China, Morgan is not wanted; Michurin is."[10] The reason for siding with Michurinism was said to be due to its "proven accomplishments," along with those of Lysenko. Compare those, said the *People's Daily*, with the useless fruit-fly experiments of the Morganists. But again, the Chinese took a different path than the USSR: although Mendelist-Morganist biology was now banned, no one would face penalties for failing to acknowledge its superiority. At least for the moment, Chinese scientists could believe what they wanted: as long as they did not teach or conduct research on classical genetics.

In June 1952, the *People's Daily* published an article titled "The Struggle to Persist in the Michurinist Direction of Biological Science."[11] The article reflected the views of the Central Committee of the CCP and reveals the extent of party support for Michurinism: "It was a great achievement that Marxism and Leninism were consciously and thoroughly applied by Michurinist biology.... Michurinist biology is the fundamental revolution of biological science." The article completely rejected genetics and called for:

> Reforming biological science thoroughly with Michurinist biology. Besides genetics, ecology, cytology, embryology, and microbiology ... are all outdated biological sciences which must be completely reformed.... One can learn Michurinist biology by criticizing outdated biology and genetics.... The courses in the departments of biology in universities should be completely reformed.[12]

Thus did pseudoscience become the policy of the Chinese state to the complete exclusion of genetics, as it had done in the USSR. Big brother and little brother were arm in arm.

In the early 1950s, with the full authority of the CCP, Chinese publications at all levels became deluged with Michurinist tracts. A new periodical, *Soviet Agricultural Science*, provided even more translations from Soviet sources. Major Chinese publishers like the esteemed *Science Press* and the *People's Education Press* went completely over to Michurinist periodicals and textbooks. Major Chinese libraries were required to remove key foreign biological journals.

Biologists who had strayed from the CCP line now began to be reprimanded in public. They were subjected to re-education training and saw their writings removed from circulation and destroyed. Many continued to accept classical genetics but dared not mention the subject in public or use it in their teaching and research. By 1952, however, in a turnaround that surprised the Chinese, in the USSR Michurinism and even Lysenko himself had begun to come under criticism. At the end of 1954, the Chinese Academy of Sciences took note of these Soviet debates and published translations of the critical articles. Mao himself learned of the open criticism of Lysenko.[13]

Just as the Chinese Academy of Sciences was beginning to criticize aspects of Michurinism, the Sino-Soviet Friendship Association prepared to celebrate the centennial of the old breeder's birth. At the opening address of the celebration in October 1955, the speaker assured the participants that the questioning of Lysenko's work was overblown and that Michurinism was on a secure footing both in the USSR and in China. This confidence was undercut in December of that year by the visit of Hans Stubbe, the president of the East German Academy of Agricultural Sciences, at the invitation of the Chinese Academy of Science. Stubbe was an antifascist who had been dismissed in 1936 by the Nazis from his post at the Kaiser Wilhelm Institute for Breeding Research. Unlike the followers of Michurinism, Stubbe had tested one of its claims and published the negative result in a scientific journal.[14] One tenet of Michurinist biology and Lysenkoism was that plants can be heritably altered by grafting, so that there is no real difference between hybridization (animal or plant breeding with an individual of another species) achieved by grafting and that from sexual hybridization. Stubbe had conducted a five-year experiment to test the claim and found "no evidence of the existence of this phenomenon." Stubbe openly accused Lysenko of incompetence. This point he made so emphatically that it was passed on to the CCP and eventually to Mao. Then in April 1956 came the shattering news that Lysenko had been removed from the presidency of the Lenin Academy. This intelligence was provided by a Soviet visitor, N. V. Tsitsin, who while in China also criticized Lysenko directly.

It is noteworthy that at the time of Stubbe's resistance, Lysenkoist pseudoscience had replaced modern biology in schools and universities in other countries behind the Iron Curtain, including Bulgaria, Hungary, Rumania, and Czechoslovakia. As geneticist Rudolf Hagemann wrote, "The damage done by Lysenkoism in those countries to teaching and research in genetics and biology was severe and long lasting."[15] It is startling to realize that at this time Michurinist pseudoscience had taken over the Soviet Union, China, and most of the other countries of the Soviet bloc, a giant territory that included hundreds of millions of people, all the while being totally and, to anyone (such as Stubbe) who took the trouble to notice, transparently false. This shows as nothing else could how ideology and pseudoscience can triumph over reason and science.

As China was preparing its second five-year plan, the open crisis in genetics raised the question of whether the nation was depending so much on Soviet scientific expertise that it was failing to develop its own. Mao, who had taken Stubbe's blunt criticisms to heart, now instructed the CCP to conduct research to ascertain whether Michurinist biology was correct and to convene a symposium on genetics before the end of 1956. At the same time, Mao announced his *Double-Hundred* slogan: "Let a hundred flowers blossom and a hundred schools of thought contend"—supposedly intended to improve the party's relations with intellectuals. After a brief period of harmony during which liberal intellectuals felt safe to reveal their true sympathies, in a giant double-cross, an "anti-rightist" campaign rounded up hundreds of thousands and sent them to concentration camps.

The Qingdao Symposium

Back in the Soviet Union, as we noted, the 20th Congress of the Communist Party began on February 14, 1956. In a closed session on February 25 came Khrushchev's four-hour "Secret Speech," in which he demolished Stalin's reputation. By this time, hundreds of Soviet scientists had signed a petition asking for Lysenko to be removed from the presidency of the Lenin Academy of Agricultural Science. In April, he stepped down. The Chinese learned of these events, and Lu Dingyi, the director of the CCP's propaganda department, noted them in a speech. Lu, a member of the Central Committee, consulted with both the geneticists and the Michurinists and announced that the ban on Mendelist-Morganist biology would be lifted. The Propaganda Department, the Chinese Academy of Sciences (CAS), and the Ministry of Higher Education produced a report laying out the history of the Lysenko affair in Russia.[16]

The symposium that Mao had ordered opened on August 10, 1956 in Qingdao in Shandong Province and lasted for two weeks. Forty-eight scientists were among the attendees, with both camps equally represented. The remaining seventy-three participants came from government, the Chinese Communist Party, and the Chinese Academy of Sciences. Yu Guangyuan, representing the Ministry of Higher Education, called for rejecting political labels such as *Morganism*. With hindsight, we can see that was essential to progress, as the example of the USSR showed that these labels had replaced thought and prejudged scientists by their ideology, in much the same way that the tags *liberal* or *conservative* stifle debate in the United States today. Yu said that disagreements in science could be settled only through "free discussion and scientific practice," another essential step. He noted that Lysenko's view that "chance [probability] is the enemy of science was philosophically wrong," thus endorsing the role of random mutations in genetics.[17]

A verbatim copy of the entire proceedings was printed and distributed nationally. The geneticists were thus able to put on the record a great deal of information about the current state of their science worldwide and its vital contributions to molecular biology, biochemistry, biophysics, and so on. They recommended that the CAS open an institute of genetics in Shanghai that would operate under the principles of modern biology, while the Michurinists would run a second branch in Beijing. In educational institutions across China, both schools of thought would be taught. The Shanghai institute was directed by C. C. Tan, who had received his doctorate under Dobzhansky at Cal Tech. The presence of the two rival institutions afforded an experiment to see which doctrine would flourish. There was no contest, as by the early 1960s Michurinist biology had virtually disappeared from the Chinese scientific literature—not surprising, since its practitioners did almost no publishable research. As the cruel saying goes, "Science advances one funeral at a time," and soon the old Michurinists were gone, to be replaced by Mao's pseudoscience.

Mao's Constitution

We can date the beginning of Lysenkoism in the USSR with Lysenko's 1929 speech and its end with the 1964 fall of Khrushchev and the statement by the president of the Soviet Academy of Sciences that Lysenko was no longer protected from criticism. Thus, Lysenkoism as official policy of the Soviet Union had a run of thirty-five years. In China, Lysenkoism got its start with the 1949 Revolution and was put to rest at the 1956 Qingdao symposium, a span of only seven years. China adopted Lysenkoism from the Soviets and then followed their example by rejecting it, but on a much more compressed timetable. But another difference lay behind the timing. Stalin believed for his entire life in the inheritance of acquired characteristics. He made this clear on many occasions: thus to deny Lysenkoism and the inheritance of acquired characteristics was to deny the thought of Stalin. Once his master died, Lysenko was on a road to Samarra, though his full rejection still took a decade. Mao adopted Michurinism, but evidently had no commitment of his own to it. He was willing to put the matter up for open debate at the Qingdao symposium—and follow where it led. Chinese scientists could question Michurinism without risking the fate of Vavilov. But that did not mean that Mao would stay out of Chinese agricultural policy: just the opposite.

One of the Soviet books the Chinese translated into their own language was *Principles of Agronomy and Soil Science*, by Soviet agronomist Vasily Williams, the son of an American railroad engineer who had immigrated to Russia in the mid-nineteenth century. Williams was one of the few Russian scientists who had supported the October Revolution, a distinction that the Bolsheviks would not forget. Williams was twice awarded the Lenin Prize, served as a deputy of the Supreme Soviet, and was a member of the Academy of Sciences. Like Lysenko, few of Williams's ideas ever worked, but this did not matter since he was right on ideology. Between 1930 and 1933, Williams's false accusations of sabotage helped lead to the downfall of a large group of agronomists.

Williams was one of Lysenko's heroes, and each enthusiastically endorsed the views of the other. In 1948, in a long letter to Stalin in which he threatened to resign, as he did on several occasions, Lysenko wrote, "I have long accepted, and I share and am developing, the Williams teaching on farming and soil development and the Michurinist doctrine of the development of organisms. Both are doctrines of a single scientific trend."[18] Williams had invented a complex system of crop rotation, tilling, and, above all, planting perennial grasses, which he claimed would preserve the moisture-retaining structure of the soil. Grain would be planted only every third year, with the land lying fallow in the two intervening ones. Medieval peasants had done the same, Williams's adherents claimed, so why not follow the time-tested practice? Other agronomists recommended fertilizers and shallow planting, which drew a familiar charge from Williams: that they were "wreckers of socialist agriculture." Williams's grasslands theory won the argument. As Khrushchev later admitted, "We were short of capital [necessary to buy fertilizer] and so Williams's theory was more attractive."[19] But when tried in Ukraine, Williams's methods brought no improvement. While still at his guerilla headquarters in Yanan in Shaanxi province of northern China, Mao had read Williams's book on soil and often quoted both him and Lysenko.

In the Great Leap Forward that began in 1958, after Michurinism had been given up officially, Mao, who had never studied science or agronomy, drew up a master plan, or "constitution," for Chinese agriculture.[20] It was mandatory for every farmer in China. The program was not directly Lysenkoist, but nevertheless drew heavily on Soviet practice and experience. It substituted Mao's pseudoscience for the best practices of agronomy. His constitution comprised eight key elements, some reasonable and others transparently false and, when carried to extremes, deadly:

1. **The popularization of new breeds and seeds.** This goal followed naturally from the work of Michurin, who had made his fame developing new plant breeds. As in the Soviet Union, where warblers were said to hatch cuckoos, extravagant claims of success immediately began to appear. Schoolchildren and their teachers had allegedly crossbred a papaya with a pumpkin. In Beijing, scientists had crossed artichokes with sunflowers, rice with corn and sorghum, and tomatoes with eggplant. The most striking of these claims was a cotton plant said to have been crossed with a tomato to produce red cotton. One worker planned to cross sorghum, corn, and sugar, so that one plant would somehow produce all three. He also intended to produce a high-yield rice using "highly advanced agricultural techniques." These miraculous results were not restricted to plants: peasants

had used artificial insemination to cross a pig with a Holstein cow. These spurious claims were given wide publicity and, like some of those below, helped to produce a false sense of confidence in Chinese agriculture that encouraged consumption rather than storage of grain.

2. **Dense planting.** Like Lysenko, Mao had become convinced of the advantages of planting seedlings close together. "With company they grow easily;" he said, "when they all grow together, they will be comfortable; a lone tree doesn't grow."[21] The claim jibed with Mao's focus on class struggle. Again like Lysenko, he believed that plants within the same "class" would not compete. In response, many communes densely planted fields of wheat, cotton, sorghum, and millet—but the crowded seedlings died. Articles in the Chinese press created just the opposite impression: that dense planting was a huge success, they claimed. One photograph showed children sitting atop wheat so dense that it supported their weight. A photographer later confessed that the children had been sitting on a bench. Most peasants failed to follow the directive to plant closely, but this apostasy had to be hidden from Mao. When he visited the countryside, rice plants from many fields would be transplanted along his route to create the impression of great density. After he left, the rice seedlings were replanted in their original location. Mao's doctor later said, "All of China was a stage, all the people performers in an extravaganza for Mao."[22]

3. **Deep ploughing.** Stalin had endorsed ploughing in deep furrows, and Mao carried this idea even further. Teams typically dug furrows by hand three to four feet deep, but some went much deeper. In Guizhou Province, peasants had to tie ropes around their waists to keep themselves from drowning in the trench water from the deep furrows they were digging. Agricultural exhibits claimed that the deeper wheat was planted, the taller the plants grew.

4. **Increased fertilization.** Throughout his career, Lysenko would repeatedly claim that some long-practiced method had been wrong all along and should immediately be replaced by his latest idea, which would quickly show its merit—although few ever

did. Immediately after the war, he focused on finding the best fertilizer. In 1946, Lysenko announced in *Izvestia*, a newspaper that promulgated state propaganda, that "all the fertilizers that we introduce into the soil, even in assimilable form, are first absorbed by the microflora, and it is the products of the microflora's life activity that provide nourishment to agricultural plants."[23] This did away with a century of scientific research which had shown that plants absorb minerals and organic molecules directly from the water in which they are dissolved. Few accepted Lysenko's claim, and it quietly disappeared. He next came up with the idea of "stretching" superphosphate fertilizer with manure and lime, delivering speech after speech touting this solution. But tests at the Lenin Hills agricultural station failed to confirm it. In fact, the lime proved downright harmful. Once again, Lysenko escaped criticism and continued to remain within Khrushchev's favor. Following the Soviet example, the Chinese government ceased investing in manufacturing plants to produce chemicals needed for fertilizer and instructed peasants to mix manure with soil in a ratio of one to nine. The peasants went further, throwing household rubbish and other trash, including broken glass, onto their fields. Another method tried was to heat and smoke soil for ten days, which supposedly would give it the qualities of manure. Like much of Mao's plan, these methods required a huge amount of peasant labor, but had little to show for it.

5. **Innovation of farm tools.** At the time of the Revolution, China had not a single factory producing tractors, and almost no one in the country had ever seen one. The first plant to build tractors did not open until 1958, at the beginning of the Great Leap Forward. It produced the large, heavy tractors favored by the Soviets for their giant collective farms, instead of the small "walking tractor" that would have been of much more use to the farmers of China. When these tractors were finally made in quantity in the 1980s, Chinese farming became much more productive.

6. **Improved field management.** As noted earlier, Mao had read Vasily Williams's book on agronomy, with its recommendation that the number of acres planted be reduced and the rest allowed

to lie fallow or serve as pasturage. Typically, Mao turned this into a slogan: "Plant less, produce more, harvest less."[24]
7. **Pest control.** The Four Pests campaign (see Figure 6.2) began in the first year of the Great Leap Forward. It aimed to eliminate rats, flies, mosquitos, and sparrows. Something like it had been part of the Great Stalin plan reviewed earlier. Sparrows were thought to consume significant amounts of grain, so their nests were destroyed, eggs smashed, and chicks killed. Millions of Chinese organized into groups that banged pots and pans to prevent the hapless sparrows from roosting, causing the poor creatures to die of exhaustion. The campaign was so successful that it drove the Eurasian tree sparrow to the verge of extinction in China. Locust and insect populations surged, leading Mao in 1960 to switch the target from sparrows to bed bugs.
8. **Increased irrigation.** The final article in Mao's constitution for Chinese agriculture was the requirement that every county in China build a dam and impound a reservoir. This produced a multitude of small reservoirs, designed and constructed by peasants. Most of the dams were made of earth and quickly collapsed. An agricultural official in the 1990s called the dam-building scheme "completely worthless." These dams required a vast amount of human labor and dislocated millions of people whose land the rising waters would inundate.

Even after the Great Leap Forward ended in 1962, Mao's faith in his own agronomy remained unshaken (see Figure 6.3). In 1964, at Dazhai in Shanxi Province, he had a working replica of his eight-point constitution built, which was visited by millions. His confidence no doubt stemmed from the faked evidence he had been shown of the success of his program. In 1958, for example, he visited a model commune in Hebei Province. His car drove through a half mile of vegetables that, Mao was told, peasants had dumped along the road because they could eat no more and did not know what to do with the surplus. When Mao got to the commune headquarters, officials told him that the peasants were eating five meals a day and that the grain harvest had quadrupled. The widespread belief in an agricultural bounty led people to eat as much as they liked—to store excess grain for future consumption

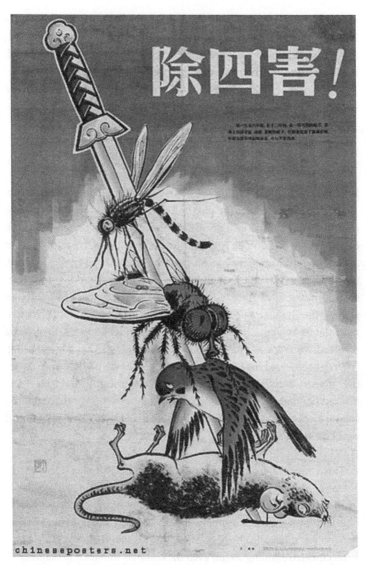

Figure 6.2 The four pests
Wikimedia Commons

would be pointless. Following the example of the big brother in the Holodomor, China's grain exports doubled, from 1958 through 1961, and imports declined. Exports to the USSR rose by half, and North Korea, North Vietnam, and Albania received free shipments of grain.[25] By the winter of 1958–1959, Chinese granaries were bare, and famine began to stalk the land. (See Figures 6.2 and 6.3.)

Tombstone

In his 2012 prize-winning book *Tombstone*, Yang Jisheng (b. 1940) poignantly brings the cost of Mao's policies down from the tens of millions of deaths to that of a single person: his father.[26] As with our focus on Vavilov, it is useful to look at a single example from among the multitude, to put one human face on the terror. In 1959, teenager Yang was a dedicated Chinese Communist and editor of his school's newspaper, *Young Communist*. In his "New Year's Message," he praised the Great Leap Forward so effectively that his school principal read the essay verbatim to the school assembly. Then in April, Yang received word from a childhood friend that his father was starving. "Hurry back," said his friend, "and take some rice if you can. Your father doesn't even have the strength to strip bark from the trees—he's starved beyond helping himself." Yang grabbed a bag of rice from the school kitchen and rushed to his hometown of Wanli. What he found shocked him:

> The elm tree in front of our house had been reduced to a barkless trunk, and even its roots had been dug up and stripped, leaving only a ragged hole in the earth. The pond was dry; neighbors said it had been drained to dredge for rank-tasting mollusks that had never been eaten in the past. There was no sound of dogs barking, no chickens running about; even the children who used to scamper through the lanes remained at home. Wanli was like a ghost town.[27]

The family's pride and joy, its water buffalo, had been slaughtered for meat, which was then confiscated by a government production team. Inside Yang's family home, not a grain of rice nor a drop of water

Figure 6.3 The people's commune is good; our great leader Chairman Mao, during the Great Leap Forward of 1958, on an inspection tour across China
Bridgeman Images

was to be found. His father lay on his bed, like a "human skeleton," too weak to go to the well. Yang made rice congee, but his father had not the strength to eat it. Three days later, he died.

Yang grieved his father's death, but was so indoctrinated that it did not occur to him to blame the government:

> I believed that what was happening in my home village was isolated, and that my father's death was merely one family's tragedy. Compared with the advent of the great Communist society, what was my family's petty misfortune? The party had taught me to sacrifice the self for the greater good when encountering difficulty, and I was completely obedient. I maintained this frame of mind right up until the Cultural Revolution.[28]

The next year, Yang entered Tsinghua University in Beijing and, while there, joined the CCP. After he graduated, he was assigned to the Xinhua News Agency. His work as a reporter gave him a much broader view and led him to realize that his father's death was not an isolated tragedy for one family, but a national one.

Yang read voluminously on the famine and traveled widely interviewing survivors. He describes how, because it was forbidden to blame the government even by implication, "tens of millions departed this world in an atmosphere of mute apathy."[29] As in the Holodomor, the government set impossibly high grain quotas and took the lion's share, leaving the people to starve. Cooking was forbidden, and people were forced to eat in communal kitchens where there was never enough. After the grain ran out, people ate wild herbs, then the bark from the trees, bird droppings, rats, corpses, and finally, members of their own family. It cannot be ignored that the two giant totalitarian states of the twentieth century each reduced their citizens to cannibalism.[30]

In early 1962, just after the Great Leap Forward ended, Liu Shaoqi, first vice chairman of the PRC, said in an official statement that 70 percent of the famine was due to manmade errors.[31] This was little more than an educated guess, since at that early time most of the data were missing or unreliable. What is significant about Liu's assessment is that a top Chinese official would lay responsibility for the death of tens of millions to human error, which had to mean by Mao and the CCP. In 1981, the Party officially recognized that the famine was mainly due to mistakes made on behalf of the Great Leap Forward. This admission paved the way for the PRC to abandon pseudoscience and return to mainstream biology.

Many scholars and demographers, foreign and Chinese, have attempted to estimate the death toll from the Great Famine. Missing information, voluminous records, and deliberate obfuscation complicate the task. Yang Jisheng estimates that thirty-six million died.[32] Historian of China Frank Dikötter relies on the work of senior party member Chen Yizi, who was a member of a large group that sifted through a raft of party documents to estimate the death toll.[33] Dikötter puts it at a minimum of forty-five million deaths. He notes that some historians believe the total could be as high as fifty to sixty million, in the same range as that from the two great wars of the twentieth century. The books of both Yang and Dikötter are banned in the PRC.

7

Jewish Physics

"We do not need them."[1]

—Hitler

In this chapter and the next, we examine the role of science denial and pseudoscience in another totalitarian state: Nazi Germany. Both Nazi antisemitism and the Holocaust stemmed from the belief that some races—however that term is defined—are inherently superior to others. The Nazis identified Germans as a *Herrenrasse*: a master race descended from an "Aryan people" that supposedly represented the pinnacle of a human racial hierarchy. This pseudoscientific belief allowed the Nazis to do as they liked with other races that were by definition subordinate to the masters. The result was the expulsion of Jews from academic positions in Germany and eventually the Holocaust, in which millions of Jews, homosexuals, disabled persons, and Romani people were deliberately murdered en masse.

Mein Kampf

As the twentieth century began, German universities were among the world leaders in science. Germany was the first country to offer the PhD degree, and when in 1887 Johns Hopkins became the first American university to follow suit, it adopted the German model. Nearly all the university's faculty members had studied in Germany. Indeed, in some fields in those days, a scientist's education could not be said to be complete without a year or two in Germany. A good example is Robert Oppenheimer, the American "father of the atom bomb." After graduating from Harvard, he studied with Nobel laureate Max Born at the University of Göttingen and while there published several papers on quantum mechanics. In Germany, Oppenheimer met most of the leading physicists of the day, some of whom would later emigrate and join him on the Manhattan Project. He received his doctorate in 1927 at age 23 and then went to the Swiss Federal Institute of Technology in Zurich to work with Austrian Nobel laureate Wolfgang Pauli before returning to America as an associate professor at the University of California at Berkeley.

It is no surprise that German scientists in the first decades of the twentieth century would take great pride both in their accomplishments and in their identity as Germans. They tended to regard science and politics as a priori incompatible and failed to recognize their own ethnic backgrounds as inherently political. As a group they were not free of antisemitism, yet German medicine and science included many Jewish professors. But this openness would soon prove incompatible with Hitler's ideology, made clear in his rabidly antisemitic *Mein Kampf*, written in 1925 while he was in prison for leading a failed coup in Munich in November 1923.

In his book, Hitler brushed aside science, writing that it was dangerous to devote too much school time to chemistry, physics, and mathematics. (Hitler had dropped out of high school without graduating.) Moreover, he regarded scholarly training in general as less important than development of sound bodies and character: "A person less scholarly educated [*gebildet*], but with a good firm character full of decisiveness and will power is worth more to a people as a whole than a physically degenerated, weak-willed, cowardly pacifistic individual," he wrote.[2] When asked in a 1931 interview whence would come the brainpower necessary to operate the Nazi state, Hitler replied that he would supply the brains and, as for the German middle class, "we will do what we like with them." When the interviewer asked about Jews, noting that among them were many capable people, including Einstein, Hitler was dismissive:

> Everything they [Jews] have created has been stolen from us. Everything that they know will be used against us. They should just go and foment their unrest among other peoples. We do not need them.[3]

These ideological beliefs were set to become state policy when on January 30, 1933, Hitler became chancellor of Germany. Hundreds of officials were quickly dismissed for being "anti-Nazi," and then on March 31 came the first openly antisemitic action, when Jewish judges in Prussia were fired. The next day a national boycott of Jewish stores took place, with Brown Shirts smashing windows and assaulting Jews. Einstein was in America at the time, and after his return to Europe, he never again set foot on German soil and resigned from the Prussian Academy of Sciences. Despite his acclaim abroad, in Germany Einstein was viewed as a traitor. The Prussian Academy followed the Führer and the Nazi party line in avowing that it had no reason to lament Einstein's absence.

The Law for the Restoration of the Career Civil Service, under which civil servants of "non-Aryan" descent were dismissed, was enacted on April 7, 1933. A non-Aryan was defined as a person with at least one Jewish parent or grandparent. This move was so bold that it surprised even the most ardent Nazi party members. On May 6 came a supplement that defined all higher education instructors as civil servants, even if they were on a private payroll, subjecting them to the new civil service laws. This stroke of the pen would devastate German universities, but especially physics, which would lose at least 25 percent of its faculty. Had Einstein remained at his German post, he would certainly have been fired as "politically undesirable," if not worse. His books were among those publicly burned, and a price was put on his head.[4]

After Einstein, the most famous German scientist of the day was likely Fritz Haber, who had developed the chemical process for using nitrogen and hydrogen gases to synthesize ammonia, an important ingredient of explosives and fertilizers. Haber also led the German production of poisonous chlorine gas for use on the World War I battlefield. For his work in producing ammonia, he received the Nobel Prize for Chemistry in 1918. Even though he was Jewish, Haber was exempt from the Civil Service law because of his service in the First World War. Nevertheless, in 1933 the Journal for All Natural Sciences (*Zeitschrift für die gesamte Naturwissenschaft*) wrote that "The founding of the Kaiser Wilhelm Institutes in Dahlem was the prelude to an influx of Jews into the physical sciences. The directorship of the Kaiser Wilhelm Institute for Physical and Electrochemistry was given to the Jew, F. Haber, the nephew of the big-time Jewish profiteer Koppel."[5] (Haber and Koppel were not related.) On April 30, Haber resigned his directorship, saying, "My tradition requires that in a scientific post, when choosing coworkers, I consider only the professional and personal characteristics of applicants, without considering their racial make-up."[6]

Another distinguished and influential German scientist of the time was Max Planck, who had discovered that energy is quantized and received the 1918 Nobel Prize in Physics for the work. In May 1933, in his capacity as president of the Kaiser Wilhelm Society, a research foundation funded by the government, Planck (who was not Jewish) arranged an audience with Hitler. In the meeting, he tried to defend Haber, only to have Hitler respond, "I have nothing at all against the Jews as such. But the Jews are all Communists and these are my enemies." When Planck urged making distinctions among Jews, Hitler proclaimed, "A Jew is a Jew." This made it impossible to argue with Hitler that the most accomplished Jewish scientists should receive favorable treatment. He is reported to have said that "if the dismissal of Jewish scientists means the annihilation of contemporary German science, then we shall do without science for a few years!"[7]

In his book, *Scientists Under Hitler*, Alan Beyerchen includes a table listing the twenty German Nobel laureates who resigned from German institutions between 1933 and 1945.[8] Eleven were physicists. All emigrated except for physicist Gustav Hertz, who was forced to resign his university position and who surrendered to the Russians at the end of the war to avoid prosecution in Germany. By one estimate, between 1933 and 1941 more than one hundred German physicists emigrated to America alone, not all of them Jewish.[9] Others went to England or to a dozen or so other countries. Who can say what these scientists might have accomplished had they instead remained in Germany, working for the war effort.

In particular, might Germany have built an atomic bomb before America? It had a head start. German scientists Otto Hahn, Lise Meitner, and Fritz Strassmann had discovered and explained atomic fission, wherein an atom of uranium, for example, splits in two. Hungarian physicist Leo Szilard, who also emigrated to America, had described the nuclear chain reaction, in which each fission event releases more than one neutron. Each excess neutron then induces a new fission event, which produces even more neutrons, and so on in an exponentially growing cascade that leads to a colossal explosion. Germany had a potential leader of a bomb project in the brilliant Werner Heisenberg (1901–1976), who won the Nobel Prize for Physics in 1932 for his work on quantum mechanics. But as the U.S. Manhattan Project showed, to produce a bomb took far more than theoretical knowledge — it took time, money, and an unwavering commitment from the very top. Hitler was counting on a short war propelled by blitzkrieg and never had the patience nor provided sufficient funds for such a long-term undertaking as building an atomic bomb, which offered no guarantee of success. Instead, he would invest heavily in the V-2 rocket, which had almost no effect on the outcome of the war.

The incessant and often inconclusive meetings of all sorts that Soviet scientists and officials were obliged to attend in the Lysenko period, one might call the overhead of politicized science. In a similar way, there was an academic overhead to Naziism, as the many Nazi meetings, rallies, and protests ate into the time that students and professors had available for their studies. To make matters worse, the number of class hours students were allowed to take was reduced to give them more time for these unproductive political activities.

German Physics

The strong nationalism evident in Germany in the early twentieth century made it natural for many German scientists to regard other countries as enemies.[10] These feelings were inflamed during the First World War with Germany on one side and England and France on the other. In the first month of the war, Germany invaded neutral Belgium, and to retaliate for civilian resistance the Germans set fire to part of the town of Leuven, including the library of its famous university. Some 300,000 books and 1,000 manuscripts were burned. In an ominous portent, Belgian civilians were herded into cattle cars and shipped to Germany. This atrocity led to a note of protest signed by eight eminent British scientists: William Bragg, William Crookes, Alexander Fleming, Horace Lamb, Oliver Lodge, William Ramsay, Lord Rayleigh, and J. J. Thomson, who would have five Nobel prizes among them. Fifteen German scientists responded, calling the British letter an "anti-German declaration formulated without any understanding of German character by English scientists ... it is advisable to remove again the unjustified English influence which has penetrated German physics."[11] They recommended that German physicists cite more German sources than English and cease publishing in English journals. German publishers should accept only articles written in German. Discoveries made by Germans should retain their German name, for example, Röntgen ray rather than X-ray.

To remain an academic statesman who could remain above the fray and hold German physics together, Planck elected to stay in his post as president of the Kaiser Wilhelm Society. But the more prominent the position, the brighter the limelight. In a 1968 interview, German physicist P. P. Ewald recalls the dilemma that public appearances by German scientists could present:

> It was on the occasion of the opening of the Kaiser Wilhelm Institute of Metals in Stuttgart, and Planck as president of the Kaiser Wilhelm Gesellschaft came to the opening. And he had to give the talk ... and this must have been in '34, and we were all ... waiting to see what he would do ... because at that time it was prescribed officially that you had to open such addresses with Heil Hitler. Well, Planck stood on the rostrum and lifted his hand half high, and let it sink again. He did it a second time. Then finally the hand came up, and he said, "Heil Hitler." I was terribly disappointed .... Looking back, now, it was the only thing you could do if you didn't want to jeopardize the whole Kaiser Wilhelm Gesellschaft.[12]

The Nazification of a Physicist

As Jewish scientists disappeared from German universities and research institutes, those scientists who remained gravitated toward the

Nazi Party, many becoming ardent members. One of the most prominent was physicist Philipp Lenard. Like Hitler, he was an Austrian who became an all-the-more-committed German nationalist. He received his doctorate in physics from Heidelberg in 1886 and worked for three years at the university before departing for England to study. He had learned English by reading Darwin's *Origin of Species*. Lenard seems to have been one of those academics whom some have described as "perpetually aggrieved." He took issue with the alleged aloofness of the British and concluded that the nation lacked its "great personalities" of the past. He stayed in England for only six months before returning to Germany to work as an assistant to the half-Jewish Heinrich Hertz in Bonn, who had discovered radio waves. Lenard's mentor at Heidelberg had also been Jewish, but his (Lenard's) later virulent antisemitism had not yet arisen. With Hertz, Lenard worked on the invisible rays that traveled from a negatively charged electrode (the cathode) to the positively charged one (the anode) in an evacuated tube. Lenard had advised Wilhelm Röntgen on the kind of vessel needed for these vacuum-tube experiments. Using Lenard's recommended tube, Röntgen went on to discover X-rays, but gave Lenard no credit for his part in the discovery, another insult not to be forgotten. Consolation arrived in 1905, when Lenard won one of the earliest Nobel prizes in physics for his work on cathode rays. The next year, J. J. Thomson of the Cavendish Laboratory at Cambridge would win the prize for his discovery that cathode rays are beams of a previously unknown subatomic particle: the electron. Lenard believed that Thomson had failed to give him proper credit for his work on the photoelectric effect (the emission of electrons when electromagnetic radiation, such as light, strikes a material— the principle behind today's solar panels.) Lenard felt so strongly about Thomson's alleged lack of professional ethics that he raised the matter in his Nobel lecture. Thus as World War I began, the easily offended Lenard was already puffed up with grievances against the British. (See Figure 7.1.)

Lenard was fifty-two years old when the war began, and like many who had been trained in the previous century, he found twentieth-century physics, which was becoming increasingly theoretical and mathematical, passing him by. To German academics like Lenard, the principal foe in the war was England, not France. An academic

Figure 7.1 Philipp Lenard in 1900
Wikimedia Commons

colleague fighting on the front lines reported that Lenard had written him saying, "We should especially beat the Englishmen, because [they] had never quoted him correctly."[13] But to the shock of most Germans, their armies were not ultimately victorious. Instead came a humiliating surrender under the Treaty of Versailles, which blamed Germany for the war and exacted crippling reparations.

In the early 1920s, Lenard began to read Hitler's speeches and seek ways to express his disappointment in the Weimar Republic that governed Germany from 1919 to Hitler's ascension in 1933. In 1923, after a naval cadet had failed in his attempt to assassinate the German finance minister at the time, Lenard sent the would-be assassin a congratulatory telegram. In 1922, when the assassination of a leading politician succeeded, the government ordered flags flown at half-mast. At his Heidelberg institute, Lenard refused.

Meanwhile, Lenard found yet another grievance in the acclaim that had been accorded Einstein for his general theory of relativity. Confirmation of the theory had come in 1919, when, during a total eclipse of the sun, Sir Arthur Eddington had found that the sun's massive gravity bent starlight, as general relativity had predicted. But the theory was hard to understand, even for physicists of the day, and seemed to run counter to everyday common sense. Moreover, relativity had no need of the *aether*, a medium that supposedly filled space and carried light waves. The famous Michelson-Morley experiment of 1887 had shown that the aether does not exist, but many scientists, including Lenard, preferred to retain it and reject relativity.

Despite this resistance, Einstein became a worldwide celebrity and used his bully pulpit to call for world peace. As we know all too well today, a scientist who takes a position that is unpopular with vociferous segments of the public can become a target for the wrath and attempted character assassination by certain groups. In Einstein's case, such a group was the Association of German Natural Scientists for the Preservation of Pure Science (*Arbeitsgemeinschaft deutscher Naturforscher zur Erhaltung reiner Wissenschaft*). In August 1920, Paul Weyland, who headed the group, organized a mass meeting in the Berlin Concert Hall to criticize relativity.[14] Einstein himself attended. Weyland said that relativity was an example of Jewish arrogance and nothing more than "scientific Dadaism," referring to the movement in art that expressed the revulsion that many felt for the senseless slaughter of the late war. Lenard associated himself with Weyland and became the most credentialed German scientist to oppose Einstein and relativity. Lenard's public statements and writings grew more strident, but he had not yet crossed the line into public antisemitism.

In July 1922, German scientists and physicians were scheduled to hold a Hundred Year Celebration at the Natural Researchers and Physicians meeting in Leipzig. It would include a session on relativity. In his forward to a book on the aether, Lenard had warned that relativity was not even a theory, merely a hypothesis which Lenard's own writings had falsified. By turning a matter of science into a personal fight, Lenard said, his opponents had exhibited a "known Jewish characteristic." This was Lenard's first openly antisemitic statement in a scientific publication, a fork in the road that would lead him to disgrace. Other similar remarks would follow in a cascade. In his 1943 unpublished autobiography, Lenard would call relativity a "Jewish fraud" which was only to be expected since "its originator Einstein was a Jew."[15]

In 1924, Hitler went to prison for his role in the Munich Beer Hall Putsch, an attempted takeover of the government of Bavaria that failed. A month later, Lenard and another antisemitic German physicist, Johannes Stark, published a newspaper article in which they lauded Hitler for having the same spirit as great scientists such as Faraday, Galileo, Kepler, and Newton. Some of the language in the Lenard-Stark article closely resembled that of *Mein Kampf*, which would not appear until after Hitler left prison a year later. The two scientists could already write like Nazis.

Lenard's antisemitism grew ever more strident. In 1929, he published *Grosse Naturforscher*, translated into English in 1933 under the title *Great Men of Science*.[16] One of the Nazi Party's racial specialists had suggested that Lenard write the book, so it was no surprise that it faithfully followed Nazi ideology. It comprised sixty-five biographical sketches of outstanding scientists but included none who had lived after World War I and who were thus Lenard's contemporaries. This allowed him to omit Einstein and Madame Curie (and all other women scientists). The book reflected Lenard's belief that these great men of science had been exclusively of Aryan-German lineage. As noted earlier, Heinrich Hertz, who had been his thesis advisor at Bonn, was half-Jewish, yet was of such stature that Lenard could not reasonably leave him out of the book. In his first edition, Lenard wrote, "Heinrich Hertz was the son of a lawyer ... partly of Jewish blood."[17] In later editions, he attributed Hertz's ability as an experimenter to his

Aryan mother and his later theoretical efforts, which Lenard frowned upon in general, to Hertz's Jewish father. This would not work to explain Einstein, Lenard's nemesis, so he had to shift gears. He claimed that $E = mc^2$ was not original with Einstein but was instead the work of a non-Jewish Austrian physicist who had died in the war.

Lenard was famous for his lectures, and in 1937-1938 published them in a four-volume set titled *Deutsche Physik* (German Physics). As Beyerchen writes, its "infamous opening lines gave the Aryan physics movement its battle cry:

> "German physics?" people will ask. I could have also said Aryan physics or physics of Nordic natured persons, physics of the reality-founders, of the truth-seekers, physics of those who have founded natural research.[18]

By this time, however, so-called Aryan physics had already triumphed, and there were virtually no non-Aryan physicists left in Germany to demonize and persecute. Lenard became science advisor to Hitler and in 1937 joined the Nazi Party, allowing him to receive its Golden Badge of Honor. When American troops overran Heidelberg in 1945, they expelled Lenard from his post as professor emeritus and removed his name from a branch of the university. As a final ignominy, Lenard's name was also expunged from another sort of monument, one that no one has ever seen. In 2005, the International Astronomical Union (IAU), having not done its homework, named a crater on the far side of the moon for Lenard. When his allegiance to the Nazis and to Aryan physics came to light, the IAU removed his name from the hidden crater.[19]

Heisenberg and *Das Schwarze Korps*

The one German scientist who was the most qualified, at least by training and expertise, to head a German nuclear program was Werner Heisenberg (1901-1976; see Figure 7.2). He became famous for his enunciation of the uncertainty principle, which held that we cannot know with accuracy both the position and the speed of a subatomic particle, such as a photon or electron. Heisenberg's fate under Nazism suggests that even if a German nuclear program had been well funded, it would have been dragged down by the vicious politics and antisemitism of the Nazis.

The Aryan physicists, led by Lenard and Stark, had rejected both relativity and quantum mechanics, and, indeed, theoretical physics in general. Einstein had also rejected quantum theory, a rare mistake, famously saying, "God does not play dice with the universe." In 1936, Heisenberg was a candidate to succeed famed physicist Arnold Sommerfeld, who had been his thesis advisor at Munich University. To the Nazis, Heisenberg was suspect, not least because of his endorsement of the "Jewish physics" of relativity. To ward off Heisenberg's appointment, the Aryan physicists for the first time employed the Gestapo and its journal, *Das Schwarze Korps* (the Black Corps) in a frontal attack. The July issue included an article titled "White Jews in Scholarship," identifying those non-Jewish scientists whose work was nevertheless "Jewish in character." Heisenberg was of doubtful loyalty, the article said, because he had endorsed an article favorable to Einstein and hired Jews at his research institute. Moreover, in 1934 Heisenberg had refused to sign a declaration of support for Hitler, who became so angered by Heisenberg's subsequent 1936 Nobel prize that in January 1937 he prohibited any German from accepting another Nobel award. Heisenberg and others like him, the Gestapo article concluded, should be made to "disappear" like the Jews.[20]

The Munich faculty nevertheless recommended Heisenberg to succeed Sommerfeld, but the Aryan physicists had their own list. The fight dragged on for four years until finally Heisenberg's mother visited a family friend, the mother of Gestapo head Heinrich Himmler, to plead her son's case. The tactic worked: in 1938 Himmler exonerated Heisenberg of the charges against him, citing the recommendation of his (Himmler's) family and saying:

> I am happy to be able to inform you that I do not approve the attack of the Schwarze Korps through its article, and that I have forbidden the appearance of a further attack against you ... I would consider it proper, however, if in the future you make a clear distinction for your

Figure 7.2 Werner Heisenberg (1901–1976)
Wikimedia Commons

listeners between the recognition of the results of scholarly research and the personal and political attitude of the researcher.[21]

Heisenberg thus won a reprieve from what could have been a one-way trip to the camps, but he would not succeed Sommerfeld. Instead, the Reich Education Ministry chose Wilhelm Müller, who specialized in aerodynamics and had never published in physics. Sommerfeld would say that his successor was the "worst thinkable" person for the post.[22] As the war neared its end, a team of Allied special forces smuggled Heisenberg from occupied German territory to prevent him from falling into the hands of the Soviets. Heisenberg was the most qualified German to lead the development of nuclear weapons, but this history makes it seem almost certain that, as a "White Jew," he would not have been given free rein to do so. (See Figure 7.2.)

We cannot quantify the cost of Nazism and antisemitism to German science and the war effort—but surely it was great. The Nazis crippled physics and other branches of science while in the Allied nations these disciplines flourished and made major contributions to the war effort. Moreover, those advances positioned the United States in particular, whose homeland escaped the war unscathed, to make a number of scientific breakthroughs in the 1950s and 1960s, some with the assistance of captured German scientists like rocket specialist Wernher von Braun.

World War II was a closer call for the Allies than many today may understand. From the time of Germany's invasion of Poland on September 1, 1939, until the Japanese attack on Pearl Harbor on December 7, 1941, Britain fought largely alone. There were moments when it might not have taken much for the anti-Churchill and antiwar faction in Britain to have forced the government to agree to a negotiated peace with Hitler. If more German physicists had remained at home instead of emigrating, might they not have contributed to a superior radar system, to name just one possibility, that would have enabled Germany to win the 1940 "Battle of Britain" and forced the British to seek peace terms?

8
House of Shutters

"The stronger must dominate and not blend with the weaker, thus sacrificing his own greatness."[1]

—Hitler

By March 1945, Nazi Germany was doomed. But still it fought on. To the east, the Red Army had paused its advance only 60 kilometers (37 miles) from Berlin to gather strength for its final assault on the German capital. The war would end two months later with the hammer and sickle flying from atop the Reichstag. The Germans prepared to defend the city, but all except the most fanatic Nazis knew they were only postponing the inevitable. To the west, the American Second Infantry Division had fought its way up the bluffs of Omaha Beach, helped to free St. Lo, taken Brest, fought in the Battle of the Bulge, and after guarding the critical Remagen Bridge for a week, crossed the Rhine on March 21. In early April, troops from the division entered the small German town of Hadamar. On April 6, American soldiers would discover the first concentration camp at Ohrdurf, an annex of Buchenwald, but those who took Hadamar were the first outsiders to encounter the fatal deployment of the Nazi concept of *Herrenrasse*, the master race. They did not find at Hadamar the barbed wire, guard towers, SS patrols (Schutzstaffel or Gestapo), and dogs of Ohrdurf and the other death camps, but what appeared instead to be a health sanitorium. But its name gave it away: this was the *NS-Tötungsanstalt Hadamar*, or Hadamar Nazi Killing Facility, known locally as the House of Shutters.[2] There some 15,000 innocent people were murdered in the secret Nazi T4 program, named innocuously for Tiergartenstrasse 4 in Berlin, from where the program was managed. In all, some 250,000 people in Germany and Austria, Jews and non-Jews alike, would be murdered under T4 because they were said to have "life unworthy of life," creating what was considered an unnecessary financial burden on the Nazi state and passing on their supposedly impure genes to mongrelize the master race.[3] In a way it was fitting that Hadamar was discovered before the first concentration camp, for it was at Hadamar that the methods used in the Holocaust were invented, tested, improved, even celebrated. Hadamar was also the object of the first mass atrocity trial held in the American occupation zone of Germany, before the Nuremberg trials had begun.[4]

The concept of the racial superiority of certain Germans and the inferiority of *Untermenschen*—Jews, Slavs, Romani, and others—was fundamental to Nazi ideology. The Nazis believed they were a "Nordic race" descended from proto-Aryans who, some suggested, might originally have come from the lost continent of Atlantis. To join the Nazi party, one had to have an Aryan certificate, requiring birth and marriage certificates to document the lack of Jewish ancestry. An SS officer had to trace his Aryan descent and racial purity back to 1750. We have reviewed the Holodomor in Ukraine, in which Stalin's policies caused the death of millions, even though that was not their direct purpose. But as Timothy Snyder points out, "Hitler, on the other hand, *planned in advance* to starve unwanted Soviet populations to death" (Italics original).[5] Hitler's stated goal was to expel, enslave, starve, and exterminate the Slavs of Central Europe, leaving only a few who would be taught simple arithmetic and to read traffic signs. Given this, is it really a surprise that the Nazis would turn on their own citizens whom they deemed unworthy of life?

The Eugenics Movement

The belief that heredity can be modified to improve the characteristics of future generations, such as intelligence, physical abilities, and mental health, is known as eugenics, combining the Greek words for good and growing. The idea of selectively breeding humans goes back to Plato and to Sparta. It was the essential tenet of Social Darwinism, the interpretation that Darwin's so-called survival of the fittest applied to human affairs as well as in nature. This concept was a convenient justification for the oppressive policies of colonialism, imperialism, and economic exploitation, as proponents argued that such actions were simply nature's way, God's design. On the scale of an individual, Social Darwinism came to be used to support forced sterilization and marriage restrictions.

In the 1880s, Francis Galton, a cousin of Charles Darwin, developed these beliefs and coined the term *eugenics* to describe them. Galton believed that favorable qualities in humans were inherited, though Darwin himself went out of his way to deny this interpretation of his theory. In *The Descent of Man*, Darwin wrote, "With savages, the weak in body or mind are soon eliminated; and those that survive commonly exhibit a vigorous state of health. We civilised men, on the other hand, do our utmost to check the process of elimination; we build asylums for the imbecile, the maimed, and the sick; we institute poor-laws; and our medical men exert their utmost skill to save the life of everyone to the last moment." We could not "check our sympathy," he continued, "without deterioration in the noblest part of our nature."[6] Francis Galton, and many others, rejected this principled sentiment.

Nazi eugenics had its roots in the work of the most prominent Darwinian scientist in Germany, Ernst Haeckel (1834–1919). In 1904, he openly approved the killing of disabled babies and admitted that he had long held this view. The next year, he helped found the German Society for Racial Hygiene (the society's euphemism for eugenics). Haeckel endorsed Darwin's work and invented recapitulation theory, which holds that the embryological development of an individual organism parallels and encapsulates the evolutionary development of its species, an idea that would be discredited by the mid-twentieth century. Hitler followed Haeckel and frequently used Darwinian terms and concepts. The only chapter of *Mein Kampf* to be published as a separate pamphlet was "Nation and Race." In it, Hitler wrote:

> The stronger must dominate and not blend with the weaker, thus sacrificing his own greatness. Only the born weakling can view this as cruel, but he after all is only a weak and limited man; for if this law did not prevail, any conceivable higher evolution of organic living beings would be unthinkable.... In the struggle for daily bread all those who are weak and sickly or less determined succumb, while the

struggle of the males for the female grants the right or opportunity to propagate only to the healthiest. And struggle is always a means for improving a species' health and power of resistance and, therefore, a cause of its higher evolution.[7]

Hitler blamed Jews for the German loss in World War I and the economic problems of the ensuing Weimar Republic. He claimed that Jews were engaged in a conspiracy to control international finance and through miscegenation were corrupting German racial purity. Hitler called Jews "parasites" and declared the need for Germans to seek living space to the east at the expense of the Slavs. He called on the German people "to assemble and preserve the most valuable racial elements... and raise them to the dominant position," writing that "all who are not of a good race are chaff." Germans should "occupy themselves not merely with the breeding of dogs, horses, and cats but also with care for the purity of their own blood." The elimination of the Jews "must necessarily be a bloody process," he wrote.[8]

For their model of eugenics, the Nazis turned to practices in the United States, particularly to California. Americans had been quick to put Galton's ideas into practice, and private foundations and the Harriman railroad empire had provided financial support. The American Breeder's Association was founded in 1903 to "investigate and report on heredity in the human race and emphasize the value of superior blood and the menace to society of inferior blood." Its members included Alexander Graham Bell, the inventor of the telephone, and Margaret Sanger, who founded the American birth control movement and who also believed that sterilization should be used in cases where persons had severe mental or physical defects. Plant breeder Luther Burbank became one of the most prominent turn-of-the-century proponents, endorsing the selective breeding of human beings, enforced sterilization, racial segregation, and reduced immigration. He titled an essay he wrote in *Century* magazine "The Training of the Human Plant."[9] Written just over one hundred years ago, it makes shocking reading today.

American states began to enact laws prohibiting those with perceived defects from marrying, and by the first decade of the twentieth century some states had passed legislation to permit forced sterilization. In 1907, Indiana approved a eugenics law that called for the involuntary sterilization of "confirmed criminals, idiots, imbeciles, and rapists." In practice, the law focused on men seen as sexually deviant, as evidenced by excessive masturbation or homosexuality.[10] In 1927, a revised law replaced the original, and before it was repealed in 1974, over 2,300 Hoosiers (people from Indiana) had been sterilized.[11] From 1907 to 1932, thirty-two states enacted eugenics laws. By 1930, women had become the principal victims, though the development of the tubal-ligation method did make sterilization less dangerous. The laws authorizing sterilization were tested in 1927 in the Supreme Court in *Buck v. Bell* and upheld. Carrie Buck was a Virginian born in 1906 to an impoverished mother. She was sent to a foster home, and at age sixteen a family member raped her. Buck's foster parents petitioned to have her confined at the Virginia State Colony for Epileptics and Feeble-Minded, though she was neither. When the case was adjudicated and reached the U.S. Supreme Court, Justice Oliver Wendell Holmes infamously wrote for the majority:

> We have seen more than once that the public welfare may call upon the best citizens for their lives. It would be strange if it could not call upon those who already sap the strength of the State for these lesser sacrifices, often not felt to be such by those concerned, to prevent our being swamped with incompetence. It is better for all the world, if instead of waiting to execute degenerate offspring for crime, or to let them starve for their imbecility, society can prevent those who are manifestly unfit from continuing their kind. The principle that sustains compulsory vaccination is broad enough to cover cutting the Fallopian tubes. Three generations of imbeciles are enough.[12]

Some consider this the worst decision in the history of the Supreme Court, though whether it was worse than the ruling in *Dred Scott* (no person of African ancestry could be a U.S. citizen) or *Plessy v. Ferguson* (racial segregation laws did not violate the U.S. Constitution as long as the facilities for each race were equal in quality) is a matter of opinion. Holmes's statement conveys the same message as in Hitler's quotation, above.

California passed its forced sterilization law in 1909, and it remained essentially unchallenged for seventy years. By 1964, 20,108 people had been forcibly sterilized in California. The early results were summed up in a book titled *Sterilization for Human Betterment: A Summary of Results of 6,000 Operations in California, 1909–1929* by Eugene S. Gosney, a wealthy California financier, and eugenicist Paul Popenoe. The Nazis would adopt the book to show that enforced sterilization was feasible and humane, even enlightened. The book's authors made the case that the carrot of positive eugenics—the encouragement of "fitter" families by various means—needed the negative stick of sterilization. "The longer the application of sterilization is postponed," they wrote, "the more difficult will it be to make a positive program of eugenics work."[13] The total number of enforced sterilizations in the United States had reached 64,000 by 1963 and an estimated 80,000 by the time these programs ended in the 1970s. A disproportionate number of the American victims were female and minority, showing that compulsory sterilization could be used for "racial cleansing" as well as for negative eugenics.

The U.S. eugenics laws were eventually repealed. Indiana and some other states created historical markers to call attention to their disgraceful role in the program; several have issued formal apologies and/or offered reparations. Yet involuntary sterilization continues to occur sporadically. In 2013, for example, the Center for Investigative Reporting disclosed that from 2006 to 2010, nearly 150 California female prison inmates had been sterilized without the required approval from the state. "The 16-year-old restriction on tubal ligations seemed to be news to prison health administrators, doctors, nurses and the contracting physicians," a state official said. "None of the doctors thought they needed permission to perform the surgery on inmates."[14]

Nazi Racial Hygiene

However much we may deplore enforced sterilization in twentieth-century and early-twenty-first-century America, it went no further. In Germany, in contrast, negative eugenics became state policy and, merging with Antisemitism, expanded until it brought the smoking chimneys of the death camps. It began there in the 1920s in the guise of concern about a potential degeneration of racial purity. These apprehensions had led to the appointment of professors of race hygiene at German universities and the establishment in 1927 of the Kaiser Wilhelm Institute of Anthropology, Human Heredity, and Eugenics, partially funded by the American Rockefeller Foundation. Although there were calls for legislation on sterilization and euthanasia, the focus remained on positive eugenics, but not for long. In 1931, Fritz Lenz, who held a chair at the Kaiser Wilhelm Institute in Berlin, recognized Hitler's potential to the movement: "Hitler is the first politician with truly wide influence who has recognized that the central mission of all politics is race hygiene and who will actively support this mission."[15] In July 1933, nearly two decades after compulsory sterilization began in the United States, the Nazis passed the first of the Nuremberg Laws: the Law for the Prevention of Hereditarily Diseased Offspring. Article I read, "Anyone who suffers from an inheritable disease may be surgically sterilized if, in the judgement of medical science, it could be expected that his descendants will suffer from serious inherited mental or physical defects."[16] Those defects included:

1. Congenital feeble-mindedness
2. Schizophrenia
3. Manic depression
4. Congenital epilepsy
5. Inheritable St. Vitus dance (Huntington's chorea)
6. Hereditary blindness
7. Hereditary deafness
8. Serious inheritable malformations
9. Chronic alcoholism

The decisions as to whom to sterilize were made by Hereditary Health Courts and could be appealed, but the chances of reversal were slim. Before sterilization began, the victims were removed from the view of society and placed in an institution. Those who have disappeared and have no one to defend them are easy to ignore and then to forget. But even after sterilization, handicapped persons were singled out as a drain on the public purse, a view illustrated by Figure 8.1.

Figure 8.1 "This person suffering from hereditary defects costs the community 60,000 Reichsmark during his lifetime. Fellow German, that is your money, too." From the Office of Racial Policies
Wikimedia Commons

In view of the cost to the state of keeping handicapped people alive and because their lives had been deemed unworthy, the logical next step was seen to be euthanasia. But whereas we use the term to refer to the humane relief of pain and suffering for those with incurable diseases, to the Nazis euthanasia was a euphemism for murder.

Hitler had recognized that whereas the German people would accept compulsory sterilization, they were not ready for the murder of their own citizens. In 1935, he told Gerhard Wagner, the head of the National Socialist German Doctors' League, that "in the event of war, [he] would take up the question of euthanasia and enforce it because "such a problem would be more easily solved" during wartime.[17] For one thing, this shows that only two years after taking power, Hitler was already planning for war. In wartime, public attention would be focused outward, beyond Germany, and the sacrifice of lives for the Fatherland more acceptable. To make the connection clear, in October 1939 Hitler signed an order backdated to September 1, the date of the invasion of Poland. It authorized his doctor, Karl Brandt, and *Reichsleiter* (National Leader) Phillip Bouhler, who had joined the Nazi party in July 1922 as member number 12, to begin the "mercy killings." Again, this shows that even while launching a world war, Hitler's mind was also on how to purify the German race through mass murder, which would remain an obsession until his death by his own hand on April 30, 1945, in his bunker underneath Berlin.[18]

To implement Hitler's decree, the Nazis created the Aktion T4 program. It was made part of the chancellery of the Führer, a sort of administrative office for Hitler, thus hiding it from sight. It was this program that the soldiers of the 2nd Infantry unwittingly came upon when they reached the *NS-Tötungsanstalt Hadamar* in March 1945.

The Nazi killing began with the murder of handicapped children. In 1938, a baby had been born to a family named Knauer. The exact nature of the baby's handicaps is not known, but it may have been blind and missing limbs. Doctors diagnosed the infant as an "idiot." The family appealed directly to Hitler to grant permission for the baby to be killed. The appeal passed through the chancellery, where Bouhler submitted the appeal to Hitler, who directed Brandt to kill Baby Knauer if the diagnosis was confirmed. It was, and the infant was murdered.[19] Hitler then directed Brandt and Bouhler to create a children's euthanasia program, which became Aktion T4.

Hadamar

In 1939, the psychiatric hospital at Hadamar was reconceived as one of six units of the T4 program. Under the direction of SS member Victor Brack, whose qualification was having been Himmler's chauffeur, the Hadamar staff began to sterilize children designated as "unfit" to reproduce, then moved on to murder both children and adults. In only the first eight months of 1941, 10,072 men, women, and children were murdered at Hadamar, roughly forty per day. Since it would have been impossible to bury that number of people, and especially since the Nazis hoped to keep the program secret, the victims were asphyxiated in a gas chamber using carbon dioxide in standard containers produced by the I. G. Farben chemical company. Their bodies were then cremated, sending thick, acrid clouds billowing over the nearby town, as shown in Figure 8.2. Buses and trains brought victims to the "hospital" each day, with the windows painted over. But no one was fooled: when local children saw the bus, they would say, "Here comes the murder-box again."[20] Once arrived, people were told to disrobe for a "medical examination" and recorded as having one of sixty fatal and incurable diseases. Doctors identified each person with labels representing one of three categories: murder, murder and remove brain for research, murder and extract gold teeth. On one summer day in 1941, to celebrate the cremation of their 10,000th victim, the staff held a party with beer.[21]

But protests began to mount. On August 13, 1941, Antonius Hilfrich, the bishop of Limburg, whose diocese included the T4 center at Hadamar, wrote to the Reich Ministry of Justice to point out that the killing of handicapped patients violated not only the biblical commandment but also prohibitions in the German penal code. Public sentiment, he wrote, was causing people to say that "Germany cannot win the war if there is yet a just God."[22] These protests led Hitler to halt the "euthanasia" actions in August 1941, but in the summer of 1942 they resumed. Eventually, nearly 15,000 German citizens of all

Figure 8.2 Smoke rising from the Hadamar furnaces
Wikimedia Commons

ages would be murdered at Hadamar. The last recorded Hadamar victim was a four-year-old boy diagnosed as mentally handicapped, murdered on May 29, 1945, three weeks after the German surrender.

The Hadamar Trials

Soon after the surrender and the occupation of Germany by the Allies, the U.S. military began to try Germans responsible for the murder of Allied prisoners of war. These were some of the first such trials in history, and the interpretation of international law that would be used in the famous Nuremberg trials had yet to be worked out. The Americans had intended to try the Hadamar doctors, nurses, and administrative staff, but found that under international law they had no jurisdiction to adjudicate Germans for murdering their fellow citizens. This would change when the Nuremberg trials allowed the broadly applicable charge of *crimes against humanity*. Until then, the United States could only prosecute crimes committed against their own service personnel and civilian nationals, as well as against their allies. Then the Army discovered among the Hadamar victims nearly 500 Soviet and Polish forced laborers, a tiny fraction of the millions of slave laborers used by the Nazis. The murder of these citizens of allied nations opened the door for American prosecutors to try seven Hadamar staff for the murder of "Eastern workers." The trials began on October 15, 1945, with Leon Jaworski as chief prosecutor. Three decades later, his would become a household name in America when he served as special prosecutor in the Watergate trials. One wonders whether Jaworski ever compared in his mind the bungling Watergate incompetents with the monstrous child murderers of Hadamar who had stared back at him from the German dock.

The six-person U.S. military tribunal sentenced the chief administrator of Hadamar and two male nurses to death by hanging (See Figure 8.3). They were executed on March 14, 1946. The only female defendant, Nurse Irmgard Huber, claimed that she had never killed patients.[23] Other staff corroborated her testimony, and she was released. Later the court discovered that she had selected patients for murder and falsified their death certificates. She was re-arrested and sentenced to twenty-five years. In early 1946, so-called euthanasia crimes were moved to German courts, and some of those tried for their crimes at Hadamar, including Huber, were retried. Eight years were added to her sentence, but in 1952 she was released, as were many ex-Nazis during the Cold War. Huber lived in Hadamar until she died in 1983. (See Figure 8.3).

Practices at Hadamar allowed the Nazis to discover several facts that they would expand into a holocaust. First, people could be deceived into allowing themselves to be transported to what was not a hospital but a killing center, and once there to enter what was not a shower but a gas chamber. Second, by using gas supplied by German chemical companies, scores of people could be killed at the same time. Third, their bodies did not have to be buried but could be cremated. Fourth, as Hitler had predicted, once war began, the Germans who lived near the crematoria would offer no objection (and would later say they knew nothing about them). Fifth, German doctors, nurses, and ordinary citizens could be depended on to volunteer to operate a killing center and murder helpless people, even children, on a mass scale.

Figure 8.3 Some of the Hadamar "nursing" staff taken by a U.S. Army photographer soon after American troops arrived. Fifth from the right is likely Irmgard Huber.
Imperial War Museums

In the example of Hadamar, we have a form of state science denial that is a combination of top-down and bottom-up. Hitler and the other leading Nazis set the policy—indeed, as we have seen, he outlined it unmistakably in *Mein Kampf.* At the next level were government bureaucrats, military personnel, concentration camp administrators, SS officers, and local collaborators in the occupied territories, including the Baltic states, Belgium, France, Hungary, the Netherlands, Poland, and Ukraine. Hundreds of thousands of people were involved directly or indirectly in the implementation of the Holocaust. Meanwhile, countless others, such as the residents of the little town of Hadamar, looked the other way.

Before we move on to Part II of this book, the question naturally arises as to why science denial took root and proved so deadly in the USSR, Red China, and Nazi Germany. In his book *Tombstone,* Jisheng Yang answers:

The basic reason why tens of millions of people in China starved to death was totalitarianism. While totalitarianism does not inevitably result in disasters on such a massive scale, it facilitates the development of extremely flawed policies and impedes their correction. Even more important is that in this kind of system, the government monopolizes all production and life-sustaining resources, so that once a calamity occurs, ordinary people have no means of saving themselves.[24]

Amartya Sen, winner of the Nobel Prize in Economics, put it this way:

China, although it was in many ways doing much better economically than India, still managed (unlike India) to have a famine, indeed the largest recorded famine in world history . . . while faulty governmental policies remained uncorrected for three full years. The policies went uncriticized because there were no opposition parties in parliament, no free press, and no multiparty elections. Indeed, it is precisely this lack of challenge that allowed the deeply defective policies to continue even though they were killing millions each year.[25]

The three totalitarian states we have studied were ruled by conscience-free dictators. There was no one who would dare tell Stalin, Mao, and Hitler that they were wrong. Then large masses of people took up their dictates; otherwise, they would not have worked. But the leaders of democracies can also surround themselves with sycophants who buffer them from reality. Democratically elected leaders can find themselves followers if large numbers of the public demand it as the price of reelection. Democracies too can suffer science denial on a broad scale.

PART II
STATE-SANCTIONED SCIENCE DENIAL IN DEMOCRACIES

9
AIDS

"Personally, I don't know anyone who has died of AIDS,"[1]
—*South African President Thabo Mbeki, 2003*

Between 2000 and 2005, more than 330,000 South Africans died because the government refused to implement an antiretroviral AIDS treatment program.[2]

Sometime near the beginning of the twentieth century, bush hunters from the West African country of Cameroon came in contact with the blood and bodily fluids of a diseased ape, likely a chimpanzee. In the process, a simian immunodeficiency virus labeled SIVcpz passed from the ape to the human hunters. Retrospective studies show that by the early 1970s the virus had spread widely, though it remained undiagnosed. Its symptoms would become known as Acquired Immunodeficiency Syndrome, or AIDS.

An Exotic New Disease

The first news account of an "exotic new disease" appeared on May 18, 1981, in the gay newspaper *New York Native*.[3,4] On June 5 of that year, in its *Morbidity and Mortality Weekly Report*, the U.S. Center for Disease Control (CDC) announced that five homosexual men in Los Angeles had received treatment for a rare lung infection. By the time of the report, two had died. On July 3, the CDC reported that a rare skin cancer called Kaposi's sarcoma, of unknown cause, had been diagnosed in twenty-six homosexual men, twenty in New York City and six in California.[5] As more cases of the lung infection and Kaposi's syndrome cropped up, the CDC established a taskforce on the disease, and by September 1982 had named it AIDS. By this time, virologists hypothesized that the cause was a retrovirus, which use ribonucleic acid (RNA) rather than DNA as their genetic material. In 1983, two research groups, one headed by Dr. Robert Gallo in the United States and the other by Dr. Luc Montagnier in France, confirmed that, indeed, a new retrovirus was the probable cause of AIDS. Soon, AIDS had escalated into a deadly global pandemic. No vaccine or cure has yet been discovered, but antiretroviral drugs can now dramatically slow the progress of the disease. A study published in *The Lancet* in 2017 found that a twenty-year-old HIV-positive adult on antiretroviral therapy in the United States or Canada has a life expectancy approaching that of the general population.[6] But early in the twenty-first century, AIDS denial and resistance to the use of antiviral drugs, most tragically in South Africa, led to a far different outcome.

The 1984 Press Conference

A signal moment, both in the understanding of AIDS and in the genesis of denialism and conspiracy theories, occurred on April 23, 1984, when Robert Gallo convened a joint press conference with Margaret Heckler, Secretary of Health and Human Resources under President Ronald Reagan.[7] Gallo announced that his laboratory had identified the retrovirus, named human immunodeficiency virus or HIV, that was responsible for AIDS.[8] At the time of the press conference, peer-reviewed scientific articles by Gallo and his colleagues on the finding were "in press"—accepted and in the publication pipeline—but had not yet appeared in print. That would happen only two weeks later. But for years, the press conference and its announcement of results that technically had not been published became a cause célèbre with AIDS deniers, who contended that the premature declaration gave what they alleged was the false claim that HIV causes AIDS an undeserved head start, preventing alternative causes from being considered. To make matters worse, Heckler then said that a vaccine for AIDS was only two years away. In a 2006 interview, she admitted that this rash announcement was a mistake and shouldered the blame.[9] She had based her remark on a conversation with Gallo, who no doubt had not intended that the forecast would be made public. But we should also note that in those early days, no one could have predicted that nearly forty years later, there would still be no vaccine for AIDS. (See Figure 9.1.)

A scientific brouhaha ensued, as Montagnier demanded credit for the discovery of the AIDS virus and Gallo conceded that the virus that he claimed to have been the first to identify had been sent to him from Montagnier's laboratory. President Reagan and French President Jacques Chirac became involved, the Federal Office of Research Integrity began to investigate Gallo for scientific misconduct, and Montagnier rather than Gallo won the Nobel Prize in Physiology or Medicine in 2008. The debacle sparked a suspicion of AIDS researchers and led some to deny that HIV even exists, while others granted its reality but argued that it is a harmless "passenger" virus.

Figure 9.1 Robert Gallo, co-discoverer of HIV and prolific researcher
Wikipedia Commons

It is no surprise that denial was the first reaction of many people to the news that they or a loved one had AIDS, especially in the beginning of the pandemic. Indeed, in her book *On Death and Dying*, Elizabeth Kubler-Ross described denial as the first stage of grief, followed in order by anger, bargaining, depression, and acceptance.[10] It would be a most fortunate person who does not recognize at least a few of those stages in reaction to bad news that they or their loved ones have received at some time or another. Most AIDS deniers have not had the disease themselves but have adopted the role for some combination of reasons of their own.

In her important book *The AIDS Conspiracy: Science Fights Back*, South African economist Nicoli Nattrass delineates four roles that HIV/AIDS deniers assume.[11] We see these and more in the other examples of state-sanctioned pseudoscience we are reviewing. Nattrass describes the hero scientist, the cultropreneur, the living icon, and the praise singer. These key actors are on the front lines, providing credence for AIDS denialism, serving as board members on each other's denialist organizations, personifying AIDS denial, and gulling political leaders into questioning the settled science that HIV causes AIDS. Other denialist movements have used the same strategies.

Denialist Roles

There is no doubt that Trofim Lysenko was the *hero scientist* of the Soviet pseudoscience named for him. He received many honors, including being named a Hero of Socialistic Labor, despite having few actual scientific achievements. In contrast, the hero scientist of AIDS denial, Dr. Peter Duesberg (b. 1936), has superb scientific credentials. Seth Kalichman, author of *Denying AIDS*, says that "confusion over whether HIV is the cause of AIDS is traceable" to Duesberg, whose claims appear "credible because of his ... indisputably impressive early career accomplishments."[12] Duesberg was born in Germany and received his PhD in chemistry from the University of Frankfurt in 1963, and then joined the faculty at the University of California at Berkeley, where he has served for the rest of his career. He was one of the first to isolate the genes that cause cancer—called *oncogenes*—as well as the retroviruses that allow cancer cells to replicate themselves. This work garnered him wide acclaim as well as election to the U.S. National Academy of Sciences in 1986, shortly after Gallo and Montagnier discovered HIV.

A few years later, Duesberg reversed himself and disavowed his own discovery, now saying that oncogenes and retroviruses do not cause cancer. He even dismissed the clearly established link between the human papillomavirus (HPV) and cervical cancer, the basis for the lifesaving "pap test." Any scientist prepared to go that far would certainly have no compunction about rejecting someone else's retroviral theory, such as that HIV causes AIDS. Duesberg endorsed an alternative that had been proposed in 1914, when genetics was still in its infancy: that mutations in chromosomes (the structures inside cell nuclei that carry DNA), rather than the genes themselves, cause cancer. He went on to conclude that retroviruses cannot induce *any* disease, implicitly denying that HIV causes AIDS. In 1988, he published an article in the Policy Forum of *Science Magazine* titled "HIV Is Not the Cause of AIDS."[13]

Duesberg first attributed AIDS to the use of recreational drugs, but that failed to explain the evidence that so many who contracted the disease were not drug users. Duesberg then claimed that AZT (azidothymidine), which had begun to be used to treat AIDS, instead actually caused the disease. When it was pointed out that many who have AIDS got the disease before AZT was used, Duesberg simply denied there were any such people.[14]

Duesberg has continued to proselytize against HIV as the cause of AIDS, making him a pariah in scientific circles. When in 2007 *Scientific American* published his article on chromosomal theory, it felt obliged to add this disclaimer:

> Editor's note: The author Peter Duesberg, a pioneering virologist, may be well known to his readers for his assertion that HIV is not the cause of AIDS. The biomedical community has roundly rebutted that claim many times. Duesberg's ideas about chromosomal abnormality as a root cause for cancer, in contrast, are controversial but are being actively investigated by mainstream science. We have therefore

asked Duesberg to explain that work here. This article is in no sense an endorsement by *Scientific American* of his AIDS theories.[15]

In 2009, Duesberg and co-authors published "HIV-AIDS Hypothesis Out of Touch with South African AIDS – A new Perspective," in the non-peer-reviewed journal *Medical Hypotheses*.[16] The *Journal of Acquired Immune Deficiency Syndromes*, which is peer reviewed, had turned down the paper, one of the reviewers stating that if the article were published elsewhere, the authors could face charges of scientific misconduct. Duesberg et al. knew of the warning, but nevertheless went on to publish in *Medical Hypotheses*. The article attempted to refute studies that had concluded that AIDS denial on the part of South African President Thabo Mbeki had led to 330,000 deaths. These were essentially statistical studies and well outside Duesberg's area of expertise. The abstract of the article in *Medical Hypotheses* concluded: "Further we call into question the claim that HIV antibody-positives would benefit from anti-HIV drugs, because these drugs are inevitably toxic and because there is as yet no proof that HIV causes AIDS."[17] The demand for proof often pops up when science is being denied. Big Tobacco maintained for decades that there was no proof that smoking causes lung cancer, despite the overwhelming scientific evidence that it did.

The misconduct accusation arose because Duesberg and one of his co-authors, David Rasnick, had failed to disclose a conflict of interest: that Rasnick had profited from the sale of ineffective vitamin pills as remedies for AIDS instead of AZT and other drugs. In response to "serious expressions of concern about the quality of this article," Elsevier, which publishes *Medical Hypotheses*, sent the paper to five reviewers, each of whom recommended rejection. Elsevier retracted the article and fired the journal's editor.[18]

In response to two letters of complaint about Duesberg's misconduct, one from Nathan Geffen of the Treatment Action Coalition in South Africa, the University of California opened a formal investigation. In 2010 it concluded there was "insufficient evidence ... to support ... disciplinary action" based on the Faculty Code of Conduct, citing Duesberg's "academic freedom."[19] Universities are reluctant to get in the middle of academic controversies, preferring to leave them to the journals and professional associations. Kalichman aptly sums up Peter Duesberg's career: "How one man could be the source of so many lives saved [from his research on oncogenes] and so many lives lost is the greatest paradox and human tragedy in this whole contorted affair."[20]

Nattrass coined the label *cultropreneur* to characterize duplicitous scoundrels like Rasnick, who deny a medical finding, such as that HIV causes AIDS, and promote denialist conspiracy theories while profiting from the sale of worthless alternative medicines. The most prominent AIDS cultropreneur is German vitamin magnate Matthias Rath, whose Rath Health Foundation asserted that antiretroviral medicines do not work and instead are a form of genocide foisted on the public by "the paid lackeys" of Big Pharma. Rath contended that his vitamins can not only reverse the progress of AIDS but also cure diabetes, cancer, and cardiovascular disease.[21] In 2005, Rath and Rasnick conducted unapproved, illegal clinical trials in South Africa, luring patients with food and money to substitute Rath's products for proven antiretroviral drugs. In one example from among many, the *Guardian* reports that "a pregnant woman newly diagnosed with HIV was visited at home by Rath Health Foundation employees and convinced to stop taking her antiretroviral medication in favour of Rath's vitamins; she died three months later."[22] The Treatment Action Coalition and the South African Medical Association successfully sued to force the Rath Foundation to stop the trials and desist from claiming that its products could cure AIDS. The Rath Foundation, claiming libel, in turn sued the *Guardian*, but lost.

As her sad prototype of the *living icon* of AIDS denialism, Nattrass chose a California entrepreneur and mother, the late Christine Maggiore. She was the embodiment of AIDS denialism, as she authored articles, helped produce a film, made public appearances, gave interviews, appeared on ABC's "20/20" and other TV shows, and was unswerving and tireless in her denial of the link between HIV and AIDS and her rejection of the use of antiretroviral drugs. In 1992, during a regular medical examination, Maggiore tested positive for HIV. But after meeting with Duesberg, she began to question her diagnosis and came to believe that her positive test had some cause other than AIDS, possibly the flu or her pregnancy. She founded an organization called Alive & Well AIDS Alternatives, and wrote and self-published a book titled *What If Everything You Thought You Knew about AIDS Was Wrong?*[23]

While pregnant, Maggiore declined to take antiretroviral drugs, then breast-fed her daughter, Eliza Jane Scovill, whom she did not have tested for HIV. Nor did Eliza Jane or her older brother Charlie receive any childhood vaccinations. When asked about her children on an Air America show, Maggiore replied, "They've never had respiratory problems, flus, intractable colds, ear infections, nothing. So, our choices, however radical they may seem, are extremely well-founded."[24] Seven weeks later, Eliza was dead. The Los Angeles County Coroner ruled her death as due to AIDS-related pneumonia. Maggiore disputed the finding and had the autopsy report reviewed, not by a qualified pathologist, but by a specialist in veterinary pathology. He reported that the death was due to the use of the common antibiotic amoxicillin. But independent pathologists dismissed his finding.[25]

On December 27, 2006, Christine Maggiore herself died, and the official cause immediately became controversial. Her death certificate ascribed her death to an AIDS-related infection, but the AIDS denialist community rejected that finding and put her death down to media-related stress or some other non-AIDS-related cause.[26]

The role of *praise singer* has proved indispensable to AIDS denialists. Recall that Lysenko first garnered wide attention when in 1929 *Pravda* sang the praises of the barefoot young scientist who had turned the winter fields of the Transcaucasus green. The newspaper would continue to promote him off and on for three decades. The *People's Daily* did the same for Lysenko in China—for a while. The media may thus provide a way for dissident scientists, who cannot get their articles published in peer-reviewed journals or who have no research to publish, to take their case directly to the public. We see this both in AIDS denialism and climate denialism. In the case of AIDS, the most prominent enabler has been American journalist Celia Farber (b. 1965). From 1987 to 1995, she promoted AIDS denialism in a monthly column in *Spin* magazine titled "Words from the Front." In a 2006 article in *Harper's* magazine, she defended Duesberg, saying that he had done nothing more than "point out that no one had yet proven that HIV is capable of causing a single disease, much less the twenty-five diseases that are now part of the clinical definition of AIDS."[27] Again, this is a familiar trope of denialism, demanding absolute proof before a scientific consensus can be accepted. In fact, there was and is an overwhelming consensus among scientists that HIV is the cause of AIDS. In 2005, over 5,000 scientists signed this statement: "The evidence that AIDS is caused by HIV-1 or HIV-2 is clear-cut, exhaustive and unambiguous, meeting the highest standards of science. The data fulfil exactly the same criteria as for other viral diseases, such as polio, measles and smallpox."[28]

The Villain and the Head of State

For the purposes of this book, with its broader coverage, we need to add two roles to Nattrass's list. The first is *the villain*—a truth-telling mainstream scientist whom denialists love to hate. For climate change deniers in America, villain number one has been retired NASA scientist James Hansen. He sounded the alarm about climate change and its likely effects at a June 1988 hearing of the U.S. Senate Committee on Energy and Natural Resources, in which he said that global warming "is happening now."[29] Hansen later maintained that NASA officials and the Bush White House had tried to squelch his work. The NASA Inspector General confirmed the claim, saying that Agency officials had indeed "reduced, marginalized or mischaracterized climate change science made available to the general public."[30] Michael Mann, inventor of the famous "hockey stick" graph of rising temperatures over time, has been another favorite villain for climate change deniers.[31]

For AIDS deniers in the United States, the obvious choice for villain has been Robert Gallo, the co-discoverer of HIV and the inventor of the blood test for the virus. Since many AIDS denialists say that the virus does not exist, they must discredit Gallo. It is a tall order because he was the most cited scientist in the world from 1980 to 1990 and ranked third in the world for scientific impact for the period 1983 to 2002.[32]

A classic example of an attack on a scientist-villain by a science denier is Andrew Brink's article "Debating AZT: The Pope of Aids" on a denialist website.[33] To defame Gallo, Brink begins with a quote from

Arthur Conan Doyle's *The Speckled Band*: "When a doctor does go wrong he is the first of criminals. He has nerve and he has knowledge." Brink continues:

> Since the case for Gallo's HIV-AIDS hypothesis is invariably pressed with calls to the authority of its famous protagonist, in the absence of scientific proof in the sense that most curious folk understand, it's as well that we know what kind of bloke we're relying on... an American scientific gangster who had committed so many crass, self-aggrandizing blunders in the previous decade, that he could not really be relied upon to tell the time correctly.

Another important role to add to Nattrass's list is that of the *head of state*. Dictators such as Stalin, Mao, and Hitler can simply decree that pseudoscience is state policy and ban rival scientific theories. But even in a democracy, an elected leader can achieve the same result by inaction, delay, delegation, and budgetary decisions. This was the strategy of Thabo Mbeki (b. 1942), who served as Nelson Mandela's deputy and succeeded him as president of South Africa, remaining in office from June 1999 to September 2008. Mbeki became involved in the activities of the African National Congress (ANC) at an early age. In 1962 he moved to England and earned a bachelor's degree in economics and a master's in economic geography, making him well aware of world poverty issues. Mbeki spent 1969 through 1971 in the Soviet Union being trained in Marxism-Leninism. On his return, he became the political secretary of the ANC and participated in the secret negotiations that led state president F. W. de Klerk to remove the ban of the ANC, which led directly to the end of apartheid and the election of Mandela. When the time came for Mandela to step down, Mbeki, his deputy president, was his hand-picked successor.

Mandela was focused on eradicating poverty in South Africa and delegated national AIDS policy to Mbeki. It focused on social and cultural issues associated with AIDS, rather than on HIV and how to prevent the spread of the disease. The South African economy had boomed throughout the 1990s, and poverty had declined in response. But the prevalence of HIV in the country—the number of persons per capita living with the virus—had nevertheless dramatically increased. When Mbeki took office in 1999, nearly one in five South Africans had HIV. This fact refuted Mbeki's contention that poverty and the conditions surrounding it were the sole cause of HIV prevalence, but that did not lead him to change his mind.

At the time Mbeki became president, fifteen years had elapsed since the discovery of HIV/AIDS and its treatment. The medical science was firmly established, and as we saw above, endorsed by an overwhelming consensus of researchers. But Mbeki elected not to accept that consensus. He was a highly intelligent and educated man who decided to research AIDS and its cause himself. Where did Mbeki go for information? To the internet, where one can find support for any number of conspiracy theories. He also had the benefit of a large file of AIDS denialism tracts that Anthony Brink, author of the quotation above, had given him. Brink later bragged that, "a lone radical activist lawyer [himself] had blocked the world's largest pharmaceutical corporation and turned South Africa's president and health minister adamantly and vocally against its popular drug."[34] Mbeki also sought information from Duesberg co-author and Mattias Rath ally David Rasnick. Mbeki rejected the claim that AIDS had originated in Africa and concluded that alternative explanations of the cause of AIDS, such as Duesberg's, had been suppressed and deserved a public hearing. In 2000, Mbeki appointed a Presidential Advisory Panel to adjudicate the matter. Half of the thirty-plus seats on the panel went to deniers such as Duesberg and Rasnick, with the other half going to mainstream scientists and clinicians. Nattrass quotes an American immunologist who summed up that the panel had "pretty well everyone on it who believes that HIV is not the cause of AIDS, and about 0.0001 per cent of those who oppose this view."[35]

The opportunity for public debate of the topic came when the International AIDS Conference convened in Durban, on South Africa's east coast, from July 9 through 14, 2000. On the eve of the conference, fearful that, with Mbeki's support, deniers would derail the meeting, a group of 5,000 scientists signed the Durban Declaration, quoted above, to preempt the deniers' claims. Among the signatories were eleven Nobel laureates. In his opening address to the conference, Mbeki essentially ignored the declaration and again laid the blame for AIDS squarely on extreme poverty.[36]

One can sometimes discover what a leader truly believes less by his own words than by the people appointed to major cabinet positions. We

Figure 9.2 Mbeki's panel
Zapiro, *Mail and Guardian*, March 2000

see this with climate change in the United States, where Donald Trump appointed climate deniers to major cabinet positions and they in turn assigned deniers to the second and third tiers in their departments. As health minister, Mbeki chose Manto Tshabalala-Msimang, who had a master's in public health. She rejected the use of antiretroviral drugs both for treating AIDS and to prevent transmission of the disease from mothers to children, saying, "In my heart I believe it is not right to hand [antiretroviral drugs] out to my people."[37] Instead, she endorsed pseudoscience: home remedies, including garlic, beetroot, and potatoes. After a 2006 AIDS conference in Toronto, sixty-five leading HIV/AIDS scientists signed a letter asking Mbeki to fire her, but he ignored them.[38] That same year, Tshabalala-Msimang fell ill and required a liver transplant. In her absence, her deputy, Nozizwe Madlala-Routledge, who accepted HIV as the cause of AIDS, filled in. Routledge took an AIDS test and spoke out in favor of antiretroviral AIDS treatment. She was banned from speaking on the topic, but then went on to (correctly) attribute the high rate of infant mortality in South Africa to HIV/AIDs. This was too much for Mbeki, and as soon as Tshabalala-Msimang returned to work, he fired Routledge on charges of "corruption."

During the COVID-19 pandemic, sad to say, many of us have had to become acquainted with the concept of *excess deaths*: the number above the deaths that would have occurred in the absence of a virus or disease. When scholars applied this method to the AIDS epidemic in South Africa, they concluded that between 2000 and 2005, more than 330,000 lives were lost "because a feasible and timely ARV treatment program was not implemented."[39] Thus, as we will see again, it is not only in totalitarian states that pseudoscience costs lives, but also when science deniers are elected to head democracies.

The annual number of deaths from AIDS-related diseases in South Africa peaked at about 283,000 in 2006 and declined to about 88,000 in 2021.[40] The decline coincided with the introduction of antiretroviral drugs. According to the World Bank, as of 2020 the top twenty-one countries ranked by adult prevalence of HIV/AIDS were all in Africa, many of them also impoverished.[41] By absolute number of cases in 2020, at 7.8 million South Africa ranked first globally, while India at 2.3 million ranked second.[42] Mbeki was not wrong to consider poverty as contributing to AIDS prevalence, but it is one factor among many that intersect and compound each other, making prevention and treatment more complex. These include the stigma that prevents people from seeking treatment, low levels of education, high-risk behaviors, poor medical services, conflict, war, and so on.

To bring this story up to date, we note that a crowning achievement of President George W. Bush's tenure was the establishment of the President's Emergency Plan for AIDS Relief (PEPFAR) in 2003.[43] This comprehensive initiative aimed to combat the global HIV/AIDS epidemic, offering antiretroviral medications and diagnostic testing to the countries most severely impacted by the disease. The program has saved an estimated twenty-five million lives worldwide. Although

Congress has reauthorized PEPFAR's funding twice since its inception, the program was notably absent from the emergency spending bill passed on September 30, 2023, designed to avert a government shutdown. This omission jeopardizes the program's future funding, which will now require passage of a separate bill. PEPFAR has come under heavy criticism from the radical right because some health organizations that fight AIDS also provide abortion services, although none of the funding goes directly for abortions. If PEPFAR becomes another victim of the current dysfunction of the U.S. government, America would retreat still further from its traditional moral leadership and millions will die unnecessarily. A more-cruel outcome is hard to imagine.

10

A Predictable Emergency

"We have it under control. It's going to be just fine."[1]
—President Donald Trump, January 22, 2020

The deadly virus first appeared in China, symptomized by respiratory problems and high fevers. Unlike the "Spanish flu" of 1918, which occurred before the age of air travel, airline passengers soon spread the new virus to the four corners of the earth. Thirty American tourists initially brought it back with them from China. The index case was a fifty-two-year-old Chicago man, but the virus spread so fast that it took only another forty-seven days for the World Health Organization (WHO) to declare a new global pandemic. The world was unprepared: in America alone, over one hundred million became infected, nearly eight million had to be hospitalized, and over half a million died.

Crimson Contagion

Though that scenario sounds all too familiar, in fact it was the script for *Crimson Contagion*, a joint-exercise simulation carried out between January and August 2019 by the Trump Administration's Department of Health and Human Services and twelve states. Its purpose was to test the capacity of government to respond to a hypothetical pandemic that originated in China. Three key assumptions of the exercise were: (1) The hypothetical disease was a strain of influenza, making it easier to handle since flu symptoms show up soon after infection. There is little asymptomatic spread and less need to test to determine who is infected. (2) The government had on hand at least thirty million doses of medicine to treat the disease. (3) When the period of home sheltering ended, those emerging would find a well-organized plan of treatment, emergency management, and healthcare. None of these assumptions held for the actual pandemic that was about to arrive.[2]

The results of the exercise were summarized in an unreleased draft dated October 2019 and marked in red capitals: "DO NOT DISTRIBUTE."[3] The report was finally released on September 16, 2020, but only in response to a Freedom of Information Act request. It described an unmitigated disaster, with such phrases as "lack of clarity ... confusion ... inconsistent and inaccurate response guidance and actions ... [lack] of a standard template to report information ... staff lack clear guidance ... inconsistent use of terminology regarding vaccine types and stockpiles...."[4] The government participants could not agree on which agency was in charge, no agency or state could provide information on what personal protective equipment (PPE) they had stored and available, there was no uniform policy on school closings, and so on. These were the very things that happened once the COVID-19 pandemic got going—and will happen again unless we learn the lessons of this most recent pandemic.

The final official report on Crimson Contagion from the HHS assistant secretary for preparedness and response was dated December 9, 2019. By then, the real COVID-19 virus was already circulating in China. The first cases in Wuhan would be announced on New Year's Eve. Crimson Contagion had become reality, and HHS officials and many others knew, or should have known, that the United States was woefully unprepared.

Crimson Contagion was one of several recent exercises designed to improve the U.S. government's response to a pandemic (a country- or worldwide occurrence of an infectious disease). The 2009 swine flu, caused by the H1N1 virus, had been the first flu pandemic in over forty years. It had led to sixty-one million infections in the United States, 274,304 hospitalizations, and 12,469 deaths.[5] The experience prompted the Obama administration to begin to plan for the inevitable next pandemic, an effort that was sped up when Ebola broke out in West Africa. Ebola is not as contagious as the flu, but it is much more lethal. A detailed report on the response to Ebola identified many now familiar problems and suggested agency-by-agency actions. One concrete result was the establishment of the Directorate of Global Health

Security and Biodefense within the National Security Council. When a new pandemic arrived, the new directorate would be able to move quickly to coordinate agency responses.

Another simulation was Event 201, held in October 2019 at a conference of government officials, academics, and corporate leaders.[6] It was the fourth pandemic simulation mounted since 2001 by the Johns Hopkins Center for Health Security. Event 201 was so named to suggest that of the 200 epidemics that occurred annually, one would turn into a pandemic. The simulation envisioned a new coronavirus emerging in Brazil, then moving from bats, to pigs, to farmers, and to passengers at a large international airport. A Johns Hopkins report on the result was titled "Pandemic Simulation Exercise Spotlights Massive Preparedness Gap."[7]

On January 13, 2017, during the transition to the Trump administration, President Obama's homeland security advisor, Lisa Monaco, held a detailed briefing for incoming cabinet officers and asked her replacement-to-be, Tom Bossert, to join her in the presentation.[8] The nominated officials who attended included John F. Kelly (to become homeland security secretary), Rex W. Tillerson (secretary of state), and Rick Perry (energy secretary). Ms. Monaco reported, "We modeled a new strain of flu in the exercise precisely because it's so communicable." She said, "There is no vaccine, and you would get issues like nursing homes being particularly vulnerable, shortages of ventilators."[9]

By the time COVID-19 arrived in early 2020, however, each of those officials had left the administration. In addition, Trump's national security adviser, John Bolton, had disbanded the NSC global health directorate. In May 2019, a CDC epidemiologist who had been embedded in China's disease control agency resigned after learning that her position would be eliminated in September of that year.

On the last day of 2019, the Municipal Health Commission of Wuhan reported an outbreak of pneumonia in the city. WHO staff were briefed on the event, and a Reuters headline, dated December 30, 2019, read, "Chinese officials investigate cause of pneumonia outbreak in Wuhan."[10] Later studies would show that the first patient to present symptoms of COVID-19 had done so weeks earlier, on December 1. In a few days, the number of cases jumped to twenty-seven, then to forty-seven. On January 6, the *New York Times* identified the probable source of the virus: the Huanan Seafood Market in Wuhan. This was plausible because the SARS virus (severe acute respiratory syndrome) had first been detected in another Chinese "wet market," where both domesticated and wild meat are sold. On that same day, fifty-nine people in Wuhan presented the pneumonia-like symptoms. On January 7, 2020, the CDC issued a Level 1 ("practice usual precautions") travel advisory for persons going to Wuhan, and on January 11 Chinese scientists identified the viral genome sequence of COVID-19 and posted it on an online genetics database. The first case outside China was reported on January 13 in Thailand. The CDC announced the first U.S. case on January 20, 2020, based on samples taken on January 18 in Washington State. Crimson Contagion had come true. What neither it nor any other of the dozens of simulations contemplated was a government led by science deniers.

"This Is a Flu"

As we saw in the case of President Mbeki, a democratically elected head of state can create what is in effect a state policy of science denial through public remarks, through those he or she appoints to leadership positions, and by endorsing pseudoscience. In this and the next two chapters, we will review the responses to COVID-19 of President Donald Trump and President Jair Bolsonaro of Brazil. Both first poohpoohed the danger from the COVID-19 virus, and then when that became untenable they scoffed at proven preventative measures, including social distancing, masking, and vaccination. They not only denied medical science but engaged in pseudoscience by proposing quack remedies, as Tshabalala-Msimang had done in South Africa.

President Trump's first response came on January 22, 2020, while he was attending the World Economic Summit in Davos, Switzerland. Asked about the U.S. case of COVID-19, President Trump replied, "It's one person coming in from China. We have it under control. It's going to be just fine."[11] This was exactly what many hoped and it assuaged, for example, even the experienced Dr. Deborah Birx, the U.S. global AIDS coordinator who would become the White House Coronavirus Response Coordinator on February 27.[12] President Trump would go on to repeatedly downplay the seriousness of the virus, creating a mindset among his large and devoted following that they need not take the virus seriously. When it came time to adopt preventive measures, many of them were primed to refuse.

The president of the United States receives a daily, top-secret press briefing on issues that have flared up in the past twenty-four hours and might threaten national security. By late January 2020, public health officials and the president himself were assuring Americans that there was little risk from the virus. But in a discussion during the briefing of January 28, Robert O'Brien, the U.S. national security adviser, expressed a different view. He told the president, "This will be the biggest national security threat you face in your presidency." His deputy, Matt Pottinger, chimed in to say that he agreed with that assessment. Pottinger had spent the four days before the meeting calling his medical contacts in China and Hong Kong. One told him, "Don't think SARS 2003, think influenza pandemic 1918."[13] Pottinger learned that despite Chinese claims, the virus was spreading not only animal to human, but human to human, making it hugely more dangerous. In another contrast to the Crimson Contagion assumptions, people without symptoms were also spreading the disease. Within China, citizens were not allowed to travel to or from Wuhan, but they could fly from Wuhan to international destinations—and carry the virus with them. Pottinger believed that this likely meant that the virus was already circulating in the United States. He urged the president to restrict travel from China to the United States, which Trump did on January 31, but it was already too late for that to make a difference.[14]

One dilemma that confronted U.S. government and health officials throughout the trajectory of COVID-19 was all-too-familiar to medical professionals: how to walk the fine line between lulling people into complacency on the one hand and unduly alarming them on the other. Almost without exception, every U.S. government spokesperson, especially President Trump, chose to minimize the risk. On February 2, on the CBS program *Face the Nation*, O'Brien said, "Right now, there's no reason for Americans to panic. This is something that is a low risk, we think, in the U.S."[15] At a press briefing by President Trump's Coronavirus Task Force on February 7, Health and Human Services Secretary Alex Azar, an attorney and businessman, said,

"The immediate risk to the American public is low at this time." Taskforce member Dr. Robert Redfield, the head of the Center for Disease Control (CDC) and a virologist, added, "The multilayer system that we've put out has got us in a mode of containment"—in other words, the virus was under control.[16] That same day, President Trump told author Robert Woodward in an interview, "It goes through air . . . That's always tougher than the touch. You don't have to touch things. Right? But the air, you just breathe the air and that's how it's passed. And so that's a very tricky one. That's a very delicate one. It's also more deadly than even your strenuous flus."[17] At this point, at least in a private conversation, Trump was taking the virus as a serious threat. But his public statements from this point on and for months afterward undercut that message.

On February 8, Dr. Anthony Fauci, the long-standing director of the National Institute of Allergy and Infectious Diseases (NIAID) and the chief medical advisor to the president, said that the odds of an American getting the virus were "minuscule." On the February 29 *Today* show, Dr. Fauci was asked whether Americans should be changing their habits in response to the virus. "No," he replied. "Right now, at this moment, there's no need to change anything that you're doing on a day-by-day basis. Right now the risk is still low, but this could change."[18] All of these statements were no doubt literally true: at the moment each was made, the risk of an individual American contracting the virus was low. That is true in the initial stages of any pandemic, no matter how bad it may eventually get. But those looking for a reason to downplay the seriousness of the virus, which would include President Trump, could ignore Dr. Fauci's last four words—"but this could change"—and hear what they wanted to hear.

While traveling in India on February 24, President Trump tweeted, "The Coronavirus is very much under control in the USA . . . CDC & World Health have been working hard and very smart. Stock Market starting to look very good to me!"[19] But on that same day, the Coronavirus Task Force headed by HHS Secretary Azar was coming to a different conclusion: that the attempt to "contain" the virus—to prevent it from spreading—was failing. Redfield relayed this information to Dr. Nancy Messonnier, director of the CDC National Center for Immunization and Respiratory Diseases. At a press briefing at the White House on February 25, she cautioned that the virus might cause severe disruption to everyday life. "It's not so much of a question of if this will happen in this country anymore," Messonnier said, "but a question of when this will happen—and how many people in this country will have severe illness." She concluded, "We are asking the American public to prepare for the expectation that this might be bad."[20] Dr. Redfield soon walked back her statement, telling a House subcommittee, "I think what Dr. Messonnier was trying to say . . . could have been done much more articulately." He added, "We're still committed to get aggressive containment, and I want the American public to know at this point that the risk is low."[21] After Messonnier's remarks, the stock market fell more than 3 percent, leading President Trump to tell Azar to fire her. Trump then announced that Vice President Pence would replace Azar as head of the Coronavirus Task Force. Messonnier, who had spoken truth to power, was not fired, but never again did she appear at a White House briefing.[22] She resigned from the CDC on May 14, 2021, and today she is dean of the School of Health at the University of North Carolina.

At that February 26th press conference, President Trump again played down the danger: "When you have 15 people—and the 15 within a couple of days is going to be down close to zero—that's a pretty good job we've done."[23] He summed up: "This is a flu. This is like a flu."[24] For the next three weeks, President Trump would continue to make statements that soft-pedaled the threat: On March 10, "Just stay calm. It will go away."[25] At a White House press briefing on March 15, at which he focused mainly on the decision of the Federal Reserve to lower interest rates, "Relax, we're doing great . . . It all will pass." Yet at the same event, Dr. Fauci, who was becoming more outspoken, warned, "The worst is ahead for us," saying the crisis had reached a "very, very critical point now." Earlier that day, Fauci would not rule out the possibility of a national lockdown of restaurants and bars, as had already happened in some European countries.[26]

The president's statements in the late winter and early spring of 2020 were without medical foundation, offering false hope and dangerously misleading—essentially lies that he made up. The experience of China in January showed that the number of cases would not miraculously fall to zero. Moreover, the virology already done in China had revealed that COVID-19 was both more contagious and much more lethal than the flu. In his book *Rage*, released in September 2020, author Bob Woodward reported that President Trump described this as a deliberate strategy. In an interview on March 9, 2020, he had told Woodward, "I intended to always play it down. I still like playing it down because I don't want to create a panic."[27] That attitude would cost lives.

Stop the Spread

It thus came as a surprise to the press and to Americans at large when, on March 16, President Trump reversed himself and announced the Slow the Spread campaign. He explained, "We've made the decision to further toughen the guidelines and blunt the infection now." For the next fifteen days, he said, students should study from home, people should avoid bars and restaurants, and groups should not exceed ten people.[28] This sound strategy came at the suggestion of Drs. Birx and Fauci, who even as the president spoke believed that the fifteen days would need to be extended by at least another fifteen. Then it turned out that the day before the president's announcement, the CDC had released guidelines limiting gatherings to fifty people, not ten. This made it seem that the CDC regarded COVID-19 as less of a threat than the president and his advisors and also that the White House and the CDC were not communicating, one of the potential flaws revealed by Crimson Contagion.

Nevertheless, state governors immediately began to mandate the steps that the president had outlined, including the ten-person limit. In a phrase that few had heard of, but that would quickly come into wide use, these measures were intended to *flatten the curve* of exponential growth that marked COVID-19, spreading it out over a longer period and lowering the peak demand, thus giving a chance for healthcare capacity to meet it.

But as would happen again, within a few days of the announcement of the fifteen-day restrictions, President Trump began to back away. In a late-night tweet on March 22, he said, "We cannot let the cure be worse than the problem itself," and continued, "At the end of the fifteen-day period, we will make a decision as to which way we want to go!"[29] That is, whether to continue the restrictions. On March 24, halfway through the fifteen days, he said that he hoped "to ease the guidelines and open things up to very large sections of our country . . . by Easter (Sunday, April 12)."[30] This shocked his medical advisors, who had come to believe that another fifteen days of restrictions, at a minimum, would be necessary. A review of the number of cases made this crystal clear. On March 1, a few days after Trump had predicted the number of U.S. cases would fall to zero, there were seventy-five cases; on March 15, the day before his Slow the Spread press conference, 5,018; and on March 31, the day the first fifteen-day period ended, 210,561.[31] By April 10, there were more than half a million U.S. cases.

At a small meeting in the White House residence early on Sunday, March 29, the day before the fifteen-day period ended, Drs. Birx and Fauci told President Trump that by the end of May, one hundred thousand to two hundred thousand Americans could die from the virus. (The cumulative number of deaths by May 31 turned out to be 109,909.) They recommended that he extend the Slow the Spread Campaign by thirty days. He agreed and announced the extension later that same day. This may have been the high-water mark of Trump's response to the pandemic. Only a few days later, as he and the response team waited to enter the White House Briefing Room, he stunned Dr. Birx by telling her, "We will never shut down the country again. Never."[32] By April 6, when the number of cases had reached 393,338, Trump announced, "There's tremendous light at the end of the tunnel."[33]

Around the country, by mid-April anti-shutdown protests had begun in a number of states. On April 17, Trump endorsed resistance to his own policies in a series of three tweets. First, "LIBERATE MINNESOTA!"; then a minute later, "LIBERATE MICHIGAN!" And shortly after that, "LIBERATE VIRGINIA, and save your great 2nd Amendment. It is under siege!"[34] Each of those states had a Democrat governor. The Second Amendment reads: "A well-regulated Militia, being necessary to the security of a free State, the right of the people to keep and bear Arms, shall not be infringed." In these tweets, Trump renounced his own Slow the Spread campaign, which suggested that he had been talked into it against his better judgment by Birx, Fauci, and others, and even appeared to encourage an armed rebellion. In October 2020, the FBI foiled a plot by a right-wing militia to kidnap Governor Gretchen Whitmer of Michigan in protest of her state lockdown.

Quackery

In March 2020, President Trump began daily televised White House press briefings on the progress of the virus. They were typically attended by members of the Coronavirus Task Force, but the president himself dominated the briefings and the others spoke only when he invited them to do so. Trump spoke mostly without notes and seemed to say whatever popped into his head.

Almost as soon as the virus arrived, scientists began to explore whether existing drugs might be effective in preventing infection and reducing the severity of symptoms. An inexpensive drug called hydroxychloroquine, used for treating malaria, showed promise in reducing the spread of the virus when tested on monkey cells grown in the lab—a long way from a living human being. On March 19, President Trump enthusiastically touted the drug, saying it "could be a gamechanger" and "if things don't go as planned, it's not going to kill anybody."[35] He added that the FDA had approved the drug to treat COVID-19, but in fact, the FDA had only authorized a study. On March 28, the FDA did issue an emergency use authorization (EUA) allowing doctors to prescribe hydroxychloroquine for patients in hospital with COVID-19, even though no data supported its use. In April, President Trump boasted that after an exposure to the virus he had been taking hydroxychloroquine preventively. When the two-week regimen he followed was over, he announced that not only had hydroxychloroquine not killed him but it had also gotten "tremendous, rave reviews."[36] The results of the FDA-approved study were published in the *New England Journal of Medicine* in August 2020. It concluded, "After high-risk or moderate-risk exposure to Covid-19 [sic], hydroxychloroquine did not prevent illness compatible with Covid-19."[37] On June 15, the FDA revoked the EUA. Given this confusing mess and President Trump's continued endorsement of hydroxychloroquine, it is no surprise that in September 2021, half of Republicans surveyed said that the drug was an effective treatment for COVID-19.[38]

Back to the televised White House press briefing on April 23, 2020, when to the surprise of everyone, out of nowhere President Trump said, "And then I see the disinfectant, where it knocks it out in a minute, one minute. And is there some way we can do that by injection inside or almost a cleaning?[39] It turned out that prior to the briefing, the president had met with the Department of Homeland Security's undersecretary for science and technology, William Bryan, who had informed him about the benefits of sunlight and disinfectants on sterilizing outdoor surfaces, such as those in parks and playgrounds. President Trump had in effect continued their discussion and his dangerous speculation in front of a national TV audience of the possibility that a common disinfectant could be ingested to counter COVID-19. Trump later said he had been joking, but the makers of the disinfectant Lysol did not think so, saying in a statement that "under no circumstance" should its products be used inside the human body. These pseudoscientific recommendations by the president likely prevented those who followed them from adopting other protective measures that would have worked.

11
Protective Measures

> "Maybe they're great, and maybe they're just good. Maybe they're not so good."[1]
> —President Donald Trump on Masks, August 13, 2020

The first six months of response to the pandemic were a bumbling combination of ignorance, incompetence, and mixed messages on the part of most involved, especially the CDC and President Trump. The result was to hoodwink Americans into rejecting the medical evidence showing that protective measures such as masks, social distancing, and vaccination work and respond instead on the basis of their political ideology.

The Spanish Flu

The conditions that prevailed during the 1918 Spanish flu pandemic were quite different. Vaccines played only a minor role. As the flu virus had not been discovered, there was no technology to manufacture vaccines, and the understanding of viral transmission and the effectiveness of masks were far less advanced than today. Back then, masks were used primarily to prevent those who already had the flu from expelling respiratory droplets and infecting others. As is the case today, studies of the effectiveness of masking during the Spanish flu gave mixed results. There were no N95 type masks that filter out at least 95 percent of airborne particles with a diameter of 0.3 microns or larger and no way to be sure whether those surveyed had actually worn their masks or had worn them properly.

Still, masking had some undeniable successes. In October 1918, San Francisco's health department made wearing a mask in public mandatory, fining those who failed to wear one. The number of new cases quickly declined, and by the end of November there were so few that the city revoked the mandate. Social distancing was also tried and naturally proved unpopular, but the experience of St. Louis and Philadelphia provided a test of its effectiveness. The St. Louis health department closed theaters, schools, and churches and banned gatherings. The city's mortality rate was about half that of Philadelphia, which held a large parade in September 1918 that led to overflowing hospitals.

The cause of the Spanish flu is known today to have been the H1N1 virus. It has not disappeared, but returned in 1977, as well as in the Swine flu of 2009. The Spanish flu is estimated to have infected one-third of the world's population of 500 million at the time. It led to an estimated twenty-five million deaths, though some experts have projected that it caused as many as forty to fifty million.[2]

Testing

When a virus can spread without symptoms, as is the case with COVID-19, the first line of defense—self-isolation until the symptoms abate—is not available. The only way to discover who has the disease is to test. Otherwise, those who have the virus but have no symptoms infect others and may see no reason to isolate themselves, to be vaccinated, to wear a mask, and so on. Contact tracing becomes impossible. Policymakers and healthcare professionals do not know where the virus is spreading most rapidly and are hampered in gauging its severity and responding. Testing permits more targeted quarantining and allows businesses to stay open longer. Finally, those without symptoms may develop them later and suffer other long-term complications. Early detection can lessen the severity of these effects.

Like almost everything else in the early weeks and months of COVID-19, testing was beset by problems. Until the very last day of February 2020, the FDA required hospitals and private laboratories to go to the trouble to obtain an *emergency use authorization* before they were allowed to develop tests. The CDC developed its own tests using the reverse transcription-polymerase chain reaction (RT-PCR), which was more accurate than the subsequent antigen tests. But the CDC test kits were contaminated and, according to a review, poorly designed.[3] They had to be abandoned. Even when reliable tests became available, the U.S. government never came up with a plan for distributing them. In August, the CDC changed its COVID-19 guidance, now saying that people without symptoms "do not necessarily need a test," even if they had been exposed. This bizarre advice was roundly criticized.

As with his other responses to the virus that we are reviewing, President Trump made matters worse. At a campaign rally on June 20, 2020, he said, "When you do testing to that extent, you're going to find more people, You're going to find more cases. So I said to my people, 'Slow the testing down, please.'"[4] Members of his staff tried to walk back this statement by saying that it was made in jest, but few days later Trump told reporters, "I don't kid. Let me just tell you. Let me make it clear." In March, he had been reluctant for passengers on a cruise ship to be evacuated and tested, saying "I like the numbers [of confirmed coronavirus cases in the United States] being where they are. I don't need to have the numbers double because of one ship that wasn't our fault."

It would be hard to think of a clearer instance in which a head of state put his own political self-interest ahead of the health and safety of the public. Americans would have been better off to do the exact opposite of what Trump recommended.

Masks

With the experience of the Spanish flu and the several pandemics that have struck since, one would have assumed that a modern twenty-first century nation like the United States would have had a stockpile of the best face masks available, especially the N95 type. These would be needed particularly to protect the frontline doctors and nurses who were caring for COVID-19 patients and were continually exposed. When Dr. Birx arrived to head the taskforce in early March 2020, however, she found to her shock that in the preceding two months, with the encouragement of the government, U.S. manufacturers had shipped millions of dollars' worth of masks and other personal protective equipment (PPE) to China.[5] Moreover, the production of new PPE had been largely outsourced to China and to Malaysia.

On February 25, HHS Secretary Azar told senators that the United States had thirty million N95 masks stockpiled but needed 270 million more. The scarcity meant that the focus in the early weeks of the pandemic had to be on getting masks to health workers. For example, on February 27, 2020, the CDC tweeted, "For the general public, CDC does not currently recommend using a facemask to protect against COVID-19. Everyday preventive actions to help slow the spread of respiratory illness are recommended."[6] Yet in other countries, masks had long been providing protection against respiratory ailments. As Dr. Birx notes, at the very least the CDC could have said, "Wear a mask. It might save your life."[7] The U.S. surgeon general, Dr. Jerome Adams, made matters worse by tweeting on February 29, "Seriously people. STOP BUYING MASKS! They are NOT effective in preventing the general public from catching #Coronavirus [Twitter hashtag] but if health care providers can't get them to care for sick patients, it puts them and our communities at risk!"[8] Adams soon corrected his statement and even demonstrated how to make your own mask, but the damage was done. When the president, the CDC, and the surgeon general begin the fight against a pandemic with false and dangerous advice, you know you are in trouble. At this point, the CDC had not yet recognized that roughly half of those infected with COVID-19 showed no symptoms yet were still contagious—so-called asymptomatic spread. This meant that everyone needed to wear a mask because unless they had access to reliable tests—and they were scarce as well—they could have COVID-19 but be unaware of it.

On April 3, President Trump announced that the CDC had changed its position and now recommended the use of face masks. But he advised, "It's going to be, really, a voluntary thing. You can do it. You don't have to do it. I'm choosing not to do it, but some people may want to do it, and that's okay. It may be good. Probably will. They're making a recommendation. It's only a recommendation."[9] This garbled statement did nothing but encourage people not to wear masks if they preferred not to. In a tweet later that day, Trump doubled down,

"Somehow sitting in the Oval Office, behind that beautiful Resolute Desk.... I think wearing a face mask as I greet presidents, prime ministers, dictators, kings, queens.... I don't see it for myself." The idea of wearing a mask to protect others was foreign to the president, as it was to a number of narcissist athletes and celebrities.

By May 21, the United States had accumulated 1,667,184 cases of COVID-19.[10] When visiting the Ford Motor plant that day, President Trump did wear a mask, but only behind closed doors because, he said, "I didn't want to give the press the pleasure of seeing it."[11] He evidently gave no thought to the positive result of setting an example for all Americans. However, when he visited wounded service members at Walter Reed Hospital on June 11, he did wear a mask. In a July 19 interview with Chris Wallace of Fox News, the president said, "I don't agree with the statement that if everybody wears a mask, everything disappears."[12] Of course, no responsible person had ever claimed that. In speaking about masks at the White House on August 13, he delivered another non-endorsement: "Maybe they're great, and maybe they're just good. Maybe they're not so good."[13] At a press event on September 7, President Trump asked Reuter's reporter Jeff Mason to remove his mask before answering a question, telling him, "You're very muffled." Mason declined, offering to speak louder.[14]

In early July, Dr. Birx received an email from Trump's son-in-law and senior advisor Jared Kushner. He was passing along a message from a former member of the transition team plugging the support that a Dr. Scott Atlas was providing for the administration's pandemic efforts and especially for the president's reopening of the economy. Dr. Atlas had become a regular on Fox News, where—now in the midsummer of 2020—he claimed that the risk from COVID-19 was low and that testing was not needed unless someone was seriously ill. Atlas had written several op-eds, one in *The Hill* headlined, "The Data Is In. Stop the Panic—And End the Total Isolation." He said, "The COVID-19 pandemic appears to be entering the containment phase."[15] Like Peter Duesberg, but for a much shorter period, Atlas played the role of hero scientist to the anti-maskers and anti-vaxxers.

Atlas is a radiologist and a senior fellow at Stanford University's conservative Hoover Institute. Though neither an epidemiologist nor an infectious disease expert, he argued that the science did not support the wearing of masks, that despite numerous studies to the contrary, children could not transmit the virus, and that the best strategy was to achieve "herd immunity," in which such a high percentage of people either are vaccinated or have immunity from having been infected that the virus is deprived of enough carriers and gradually fades to manageable level. For COVID-19, that percentage was unknown, but estimates ran above 90 percent. Dr. Tom Frieden, former CDC director, responded, "Herd immunity means 1 million dead Americans. That's what it would take to get to herd immunity. That's not a plan—that's a catastrophe."[16] Sweden had tried the strategy, mainly at the urging of a single specialist. The Swedish government allowed gyms, shops, schools, playgrounds, and restaurants to stay open, while urging social distancing and better hygiene for the aged and others who were most vulnerable. According to a 2020 report in the *New York Times*, this "soft lockdown" strategy had "calamitous results," as thousands more died than in other Scandinavian countries and still Sweden never achieved herd immunity.[17]

When President Trump resumed his coronavirus news conferences in July, Atlas helped prepare the briefing materials. As a *New York Times* article put it, "It was [Atlas's] ideas that spilled from the president's mouth."[18] On August 10, Atlas took up a formal position as a presidential advisor, and Drs. Birx and Fauci began to take a backseat. Atlas would resign on November 30, 2020, after the election.

In testimony on June 23, 2022, before the congressional committee investigating the pandemic response, Dr. Birx reported that the Trump White House had concealed and censored pandemic reports sent to state and local officials, accepted misinformation from Atlas, forced her to alter or delete mitigation measures from reports, and lost interest in the pandemic response in mid-2020.

Throughout the course of the pandemic, a critical question that repeatedly arose was whether masks work. This is a harder question to answer than with vaccination, where there is an official record of who received the jab. As we all know from recent experience, people may have a mask but choose not to wear it or do so improperly. It is a rare person who cannot remember, at some point in the COVID-19 pandemic, lowering their mask to eat or drink in the presence of others. This makes it difficult to study the efficacy of masks on a large scale.

But the evidence we have is clear: masks work. In one study, when mask use increased from 10 to 40 percent in Bangladeshi villages, the number of COVID-19 cases dropped by 11 percent, with a 35 percent reduction for people over age sixty. In Denmark, masks were randomly distributed, and about half the participants in the study wore them properly. This resulted in a 14 percent reduction in infections. At Mass General Brigham hospital in early 2020, before mask mandates were introduced, the infection rate among health care workers doubled every 3.6 days and rose to 21.3 percent. After universal masking was required, the rate stopped increasing, and then quickly declined to 11.4 percent.[19]

Vaccines

When the CDC and others eventually urged in the spring of 2020 that people wear masks to prevent COVID-19 infection, the recommendation did not meet already organized opposition. In contrast, by the time the COVID-19 vaccines began to be widely available in early 2021, many people around the world, including Americans, had opposed vaccination for years or decades, especially for their children, this despite an overwhelming consensus among scientists and medical professionals that vaccination is safe and effective and the absence of any evidence to the contrary. The World Health Organization calls "vaccine hesitancy (delay in acceptance or refusal)" one of the top ten threats to global health.

Vaccine denial was given a deadly boost in 1998 when the medical journal *The Lancet* published an article by U.K. physician Andrew Wakefield that reported on twelve children, most of whom, he claimed, showed evidence of being on the autism spectrum soon after they received the MMR vaccine for measles, mumps, and rubella (German measles). The claim that "the MMR vaccine causes autism" became the basis for widespread vaccine denial in the United Kingdom and United States, led by celebrities such as actress Jenny McCarthy, who also endorsed a quack remedy for autism. Wakefield turned out to have falsified his evidence and to have conducted an "elaborate fraud," according to the *British Medical Journal*.[20] An investigation found that he stood to gain up to $43 million for selling test kits for a condition that he argued affected autistic children, dubbed *autistic enterocolitis*. Wakefield thus personifies Nattrass's cultropreneur, violating his Hippocratic oath (do no harm) for the sake of his pocketbook.

The Lancet retracted Wakefield's 1998 paper and he was debarred from practicing medicine in the United Kingdom. Of course, to conspiracy theorists, this treatment of their hero scientist only revealed the existence of a medical "deep state" out to suppress the truth. Wakefield moved to the United States and continued to deny any fraud. In 2016, he directed a film titled "Vaxxed: From Cover-Up to Catastrophe." But the damage was done, and more and more American parents filed a "personal belief exemption" from vaccination for their children. A 2014 article in *The Atlantic* was titled "Wealthy L.A. Schools' Vaccination Rates Are as Low as South Sudan's."[21] Measles had virtually disappeared in the United States, but the rise in the number of cases in Washington State in 2019 led Governor Jay Inslee to declare a state of emergency and the state legislature to introduce a bill to eliminate the personal-belief exemption. It passed and went into effect in July of that year.

Vaccine denial has long been around to some degree and has always had at least a slight political overlay, as shown in Table 11.1 from Gallup.[22]

In 1954, Democrats were two percentage points more likely than Republicans to get the Salk vaccine for polio. With each subsequent

Table 11.1 Americans' Vaccine Willingness by Party

	Democrats (%)	Independents	Republicans	Gap
1954 (Polio)	61	55	59	2
1957 (Asian flu)	70	60	62	8
2002 (Smallpox)	60	56	49	11
2009 (Swine flu)	62	56	47	15
2020 (COVID-19)	81	59	47	34

pandemic, the gap in vaccine willingness widened. In 2020, the percentage of Democrats who said they would take a COVID-19 vaccine took a large jump, while Republican willingness held to about the same level as with smallpox and the swine flu. As vaccination hesitancy rose among Republicans, their confidence in science and scientists fell. In the 1970s and early 1980s, Republicans reported more confidence in the scientific community than did Democrats.[23] By the mid-1980s, Republican leaders had begun to openly question scientific findings on such topics as acid rain and the ozone hole, both due to pollution and both of which might require increased government regulation. As the evidence for anthropogenic global warming grew in the 1990s, Republican science denial escalated to outright ridicule and disdain. Republicans in Congress became virtually unanimous in their opposition, to the point that today the subject of climate change is beneath discussion to them. If you distrust a virtually unanimous conclusion by scientists that humans are responsible for climate change, then you must distrust scientists themselves. According to a survey conducted between December 1, 2020, and May 3, 2021, 64 percent of Democrats said they have a great deal of confidence in the scientific community, as compared with only 34 percent of Republicans—nearly two to one.[24] Given this lack of confidence, it is no surprise that when COVID-19 arrived Republicans were already primed to doubt medical advice. This built until Doctors Birx and Fauci suffered personal attacks and even death threats, with Dr. Fauci requiring a security detail for his protection. This exemplary public servant became the villain of the story to the deniers, as had his colleague Robert Gallo to the AIDS deniers.

Operation Warp Speed

As soon as Chinese scientists published the COVID-19 genome, the effort to produce a vaccine began. Based on past experience, it appeared the process would take several years. Unlike in the Crimson Contagion exercise, the COVID-19 vaccine had to be created from scratch, and then be approved by the FDA and manufactured in hundreds of millions of doses, distributed nationwide, and so on. The vaccine for the human papilloma virus (HPV) had taken twenty-five years to develop; hepatitis B, sixteen years; and measles, ten years. If a vaccine for COVID-19 took even half as long as measles, say, the global death toll could rival or exceed that of the Spanish flu. Instead, in what President Trump would later call a miracle, the first vaccine for COVID-19, from Moderna, became available on December 18, 2020.

This speedy and life-saving result was due to Operation Warp Speed, a public-private partnership announced by President Trump on May 15, 2020, to "produce and deliver 300 million doses of safe and effective vaccines with the initial doses available by January 2021."[25] One reason for the success of the program was its flexibility. It did not try to pick a winner in advance, but rather funded a number of different research programs, not only by American companies but also including, for example, the University of Oxford-AstraZeneca vaccine candidate. A total of $18 billion was spent on the operation, and never was taxpayer money put to better use.

On March 21, President Trump endorsed the shot, saying, "I would recommend it, and I would recommend it to a lot of people that don't want to get it, and a lot of those people voted for me, frankly." He added, "But we have our freedoms and we have to live by that, and I agree with that also."[26] He and Melania Trump had been quietly vaccinated in January before they left the White House. In an April 2021 interview, President Trump said that the Biden administration had asked him and the First Lady to record a public service announcement (PSA) recommending that people be vaccinated for COVID-19 and that he was considering doing so. When the same two reporters interviewed Trump again in November 2021 and asked why he had not made the ad, he replied, "They have not asked me."[27] No PSA was ever made. Thus, we might think, it is impossible to know whether, had the Trumps made the PSA, it would have made any difference. Given the depth of the partisan divide that had already developed over vaccination and that there had been a strong anti-vax sentiment in the country even before the pandemic, it is reasonable to assume that it might not have.

A group of clever scientists figured out a way to find out.[28] They took clips of statements that President Trump had made on Fox News about vaccination and wove them into a twenty-seven-second mock

PSA. They then used Google's algorithms to attach the ad to selected YouTube channels and posts.[29] The Fox News Channel received the most ads, 200,000 in all. The ads played just before the shows of various Fox News and other conservative commentators. Being aware that many viewers might turn off the video as soon as they realized it was about vaccines, in the first three seconds a Fox News anchor says (and with captions), "Donald Trump urges all Americans to get the COVID-19 vaccine."

The mock PSA ran in a randomized trial in 1,014 U.S. counties in forty-three states (out of a total of about 3,000 counties), for a cost of about $100,000. The ads reached nearly six million unique viewers. The number of vaccines in the average county in the test rose by 103, for a total increase of 104,036. The authors estimate that their campaign resulted in 839 avoided deaths, costing $115 per life saved, a minute fraction of the amount spent per COVID-19 patient. They conclude, "We find that a problem with political origins also has a political remedy." But because the Trumps never got around to making the PSA, that remedy was never tried on a broad, nationwide scale. To make the PSA would have taken only a minuscule amount of their time and surely would have saved many lives.

Medical science has shown that the four most effective methods for combatting the spread of a pandemic are testing, social distancing, masks, and vaccines. President Trump tried to reduce the number of COVID-19 tests, sided with and encouraged armed shutdown protesters, derided the use of masks and rarely wore one, and eventually renounced his one success: vaccination.

In the 2019 Global Health Survey conducted by Johns Hopkins University, before the pandemic had begun, the United States ranked first among all nations in its preparedness for epidemics and pandemics.[30] As of early October 2023, at 109 million the United States ranked first in total COVID-19 cases while India ranked second at 45 million.[31] At 3,518 COVID-19 deaths per million population, the United States ranks fifteenth from the top (worst) among some 200 nations. Compare that rate with South Korea at 700, Taiwan at 796, and Vietnam at 437.

12
Politicization of COVID-19 Denial

"It's a little flu. The worst has passed."[1]
—*Brazilian President Jair Bolsonaro, March 24, 2020*

During the first presidential debate on September 29, 2020, Donald Trump derided Joe Biden for always wearing a mask in public, saying, "I don't wear a mask like him [Biden]. Every time you see him, he's got a mask. He could be speaking 200 feet away . . . and he shows up with the biggest mask I've ever seen."[2] That first debate was watched by more than seventy-three million, each of whom heard the president of the United States mock and ridicule the easiest and most effective way available at the time to protect Americans from COVID-19. Two days after the debate, President Trump and First Lady Melania tested positive for coronavirus. The president then developed symptoms severe enough to send him to Walter Reed Hospital. According to Chief of Staff Mark Meadows, the president's oxygen levels dropped to a dangerous low, and his doctors considered putting him on a ventilator. His care included an antibody "cocktail" not yet approved by the FDA for other patients and a steroid injection reserved for severe cases. While still in the hospital, the president tweeted, "Don't let it dominate you. Don't be afraid of it. You're going to beat it." He went on, "We have the best medical equipment, we have the best medicines—all developed recently."[3] What he did not say was that the average American hospitalized with COVID-19 would not receive anything like the special treatment that probably saved the president's life.

When Trump returned to the White House on October 5, he stepped up to a microphone on the south portico of the White House, yanked aside his mask, and went inside, even though he was probably still contagious and was visibly struggling to breathe.[4] By that day, 220,368

Americans in total had died from COVID-19. Trump even suggested that he had contracted the virus on purpose as an act of leaderly self-sacrifice: "As your leader, I had to do that," Trump said on his return, "I knew there's danger to it—but I had to do it. I stood out front. I led. Nobody that's a leader would not do what I did. And I know there's a risk, there's a danger—but that's okay."[5]

Trump's final campaign event in 2020 was a huge rally held early on election day in Grand Rapids, Michigan. A study released as a working paper by Stanford University economists three days before the election had concluded that eighteen prior Trump rallies—from June 30, 2020, to September 12, 2020—had caused 30,000 more additional cases of COVID-19 and likely 700 additional deaths.[6] They arrived at this estimate not by tracing individual cases of rally attendees, which would have been virtually impossible, but by comparing the data for counties in which the rallies occurred with 200 counties with like demographics and COVID-19 experience, but that had no rally. The first of the eighteen rallies had taken place in Tulsa, where public health officials said that the subsequent surge in cases likely resulted from the rally. After the Stanford study was completed, the president held some three dozen more rallies around the country. At the final rally in Grand Rapids, as at all the others, neither President Trump nor Vice President Pence wore a mask, nor did most of the attendees, who stood shoulder to shoulder to cheer on their idol.

The Stanford study proved controversial, but there is no doubt that Trump's rallies and similar events, whether indoors or outdoors, caused tens of thousands of new COVID-19 cases. A much larger, though impossible to quantify, number of cases derived from the refusal of many Americans, following their president's lead, to wear masks.

Red States vs. Blue States

As COVID vaccines were becoming widely available in the United States in early 2021, Americans were polled as to whether they planned to take the shot. A survey published on January 22, 2021, found that about half of respondents had either already received a vaccination or said they would get one.⁷ A CBS poll in mid-March 2021 found roughly the same result, but also revealed a partisan split. Of Democrat respondents, 71 percent either already had the shot or said they would get it, versus only 47 percent of Republicans.⁸ In a study published in July 2022, researchers found that even when controlling for age, education, and gender— party affiliation alone predicted vaccination willingness. Republicans were significantly less likely than Democrats to have had the shot, to be willing to have it, and to recommend vaccination to a friend.⁹ In view of these findings, one would also predict that more Republicans than Democrats would have contracted COVID-19 and died from it. In this era of big data, we can test this sad prediction.

According to the Johns Hopkins University Data Center, by June 30, 2022, 87,838,623 Americans had contracted the virus and 1,017,846 had died.¹⁰ (Early in the pandemic, future presidential advisor Scott Atlas had estimated the eventual death toll at 10,000.)¹¹ A study by the Brown School of Public Health estimated that of the 641,305 COVID-19 deaths between January 2021 and April 2022, 318,000, just under half, were persons who had access to vaccines but had chosen not to be vaccinated.¹² Every second American who died during that sixteen-month period could have been saved by vaccination.

Deaths from COVID-19 in the United States occurred in a series of waves, as shown in Figure 12.1.

During the first wave in the spring of 2020, urban counties in the northeast had much higher death rates (deaths per hundred thousand residents) than less-populous counties elsewhere.¹³ But in the subsequent waves, this pattern reversed, and the largest share of deaths came from areas with lower populations. Overall, the COVID-19 death rate during the pandemic was slightly higher in less-populated regions. This seems hard to explain, until we examine the data by party affiliation.

In the spring of 2020, counties that Joe Biden won had far more deaths than the counties that Trump won. Again, this was because the pro-Biden counties were predominantly in the urban northeast. Not just because these areas lean Democrat, but more likely due to such factors as population density, to where the virus first happened to take hold, and to the number of foreign visitors. By the third wave in the winter of 2020–2021, this pattern had reversed, and pro-Trump

Two years of coronavirus deaths in the United States
Average number of daily reported coronavirus deaths in the U.S.

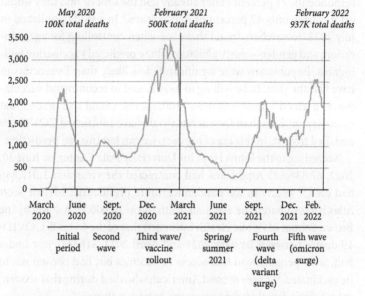

Figure 12.1 Pattern of U.S. COVID-19 Deaths
Pew Research Center[a]

[a] Bradley Jones, "The Changing Political Geography of COVID-19 Over the Last Two Years," Pew Research Center - U.S. Politics & Policy, March 3, 2022, https://www.pewresearch.org/politics/2022/03/03/the-changing-political-geography-of-covid-19-over-the-last-two-years/

counties were suffering substantially more deaths than pro-Biden counties. By the fall of 2021, death rates in pro-Trump counties were about four times greater than in counties that had supported Biden. As the Pew report sums up:

> The cumulative impact of these divergent death rates is a wide difference in total deaths from COVID-19 between the most pro-Trump

and most pro-Biden parts of the country. Since the pandemic began, counties representing the 20% of the population where Trump ran up his highest margins in 2020 have experienced nearly 70,000 more deaths from COVID-19 than have the counties representing the 20% of population where Biden performed best. Overall, the COVID-19 death rate in all counties Trump won in 2020 is substantially higher than it is in counties Biden won (as of the end of February 2022, 326 per 100,000 in Trump counties and 258 per 100,000 in Biden counties).[14]

Another study examined the excess death rate by party, pre- and post-vaccine availability.[15] Before vaccines became widely available, there was virtually no difference in excess deaths between Democrat and Republican counties. After the vaccines arrived, excess deaths among Republicans were significantly higher, the price of self-inflicted science denial.

Falling Confidence in Medicine

An obvious question is whether the disparagement and distrust of medical professionals during COVID-19 will turn into a lasting decline in confidence in medicine. The evidence is that it already has. In 1966, seven of ten Americans said they had great confidence in "the people in charge of running medicine."[16] By 2012, this had fallen to three in ten. But throughout this period up to 2018, there was little difference between the respondents according to political party: confidence in medicine fell in both groups. But by 2021, it had risen to 46 percent among Democrats, while among Republicans it had fallen to 32 percent.

That Republicans were less likely to get the COVID-19 vaccination than Democrats has led to worry the that they might decline shots for other diseases as well. Sadly, this concern also appears to have been well-grounded. As shown in Figure 12.2, the percentage of Republicans who said they got a flu shot has fallen over the last three years, while the percentage for Democrats has risen slightly.[17]

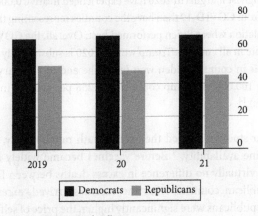

Figure 12.2 Partisan Split on Flu Vaccination
The Economist

Finally, a study published by the Commonwealth Fund on December 13, 2022, testified to the value of vaccination. The COVID-19 vaccination program in the United States prevented nearly 120 million infections, kept 18.5 million people out of the hospital, saved more than 3.25 million lives, and prevented $1.15 trillion in additional medical costs that would otherwise have been incurred.[18] What more could one ask of medicine?

Lost Window of Opportunity

When we Americans think back over the course of the virus, we tend to focus on the red-blue polarization that cost hundreds of thousands of lives. But as science author David Wallace-Wells points out, extreme partisanship around protective measures did not really begin until the fall of 2020.[19] Up to that time, there was a chance it could have been at least partly avoided. By late March and early April 2020, for example, forty states had issued stay-at-home requirements. They lasted about two weeks longer under the twenty-three Democratic governors than the seventeen Republicans, which Wallace-Wells calls a "real but small" difference. By September 2020, only four states had no recommendations or restrictions on businesses. The remaining states were more or less evenly divided politically, but the level of restriction did not track well with partisanship. As of August 2020, the West Coast states had restricted social gatherings, but in the rest of the country there was no clear difference based on partisanship. Or consider masking. By September 2020, every state had adopted some type of masking policy—again the differences were small. Down the West Coast, across the South, and up the Atlantic Coast, masking policies varied little. An August survey found that 92 percent of Democrats and 76 percent of Republicans said that they had worn a mask in stores and businesses all or most of the time. Significant, but not large. Of course, the *rhetoric* from Red-state governors and local leaders differed from those of blue states, but as Wallace-Wells notes, "Almost everyone was well within a standard deviation of what everyone else was doing."

School closings were understandably an emotional and divisive issue. But here again, the similarities exceeded the differences. In spring 2020, all but nine states closed their schools for the rest of the academic year. Three of the nine had Democratic governors, while four Republican-controlled state governments recommended closing and two left the matter to the local level. When the new school year began in the fall of 2020, red and blue states had roughly similar patterns of in-person, hybrid, and remote learning.

What of vaccination, which may have been the most divisive issue of all? The first vaccines did not become widely available until early January 2021, but in a Gallup poll conducted in mid-to-late September 2020, 54 percent of Democrats and 49 percent of Republicans said they would have the vaccination when it became available.[20] Within the ± 3 percent error of such polls, these responses were virtually indistinguishable.

Despite President Trump's downplaying of the virus and his undercutting of protective measures, as of Labor Day 2020, the Republicans and Democrats were not that far apart on measures to protect against COVID-19. Had the president followed the science and supported those measures, as he would do temporarily for vaccines, the pandemic response might not have become such a partisan issue. But now comes the issue of timing. Through September and October, as election day (November 3, 2020) grew ever closer, partisanship rose. Democrats ramped up their criticism of the Trump administration's poor response to the pandemic, while Republicans increasingly defended it, driving the two sides farther apart. In a poll that closed at the end of October, the percentage of Republicans who said they would get the jab remained at 49 percent, but for Democrats it had risen to 69 percent. It is a tempting thought that had the election not occurred until after the vaccines became available, perhaps the extreme partisanship would not have gotten as bad as it did.

To sum up, President Trump deserves criticism for downplaying the seriousness of COVID-19 from its first appearance in the United States in January 2020 until his "slow the spread" announcement in mid-March, and then again when he tweeted as he lay in his hospital bed on October 5, "Don't let it dominate you. Don't be afraid of it. You're going to beat it." He refused to wear a mask except on special occasions and routinely ridiculed those who did so. These actions caused the deaths of thousands of Americans. But Trump deserves credit for starting Operation Warp Speed, for eventually advising Americans to get the vaccine, and for making it known that he and his wife had done so. But here his advice met the existing anti-vaccination movement, which proved a more powerful influencer than fealty to the head of state, even one with the loyal base of Donald Trump. In the age of the internet, rampant conspiracy theories, and declining respect for science and for government, mass movements can establish science denial as effective national policy even without the support of a legislature or head of state. Moreover, by February 2023, Trump was disavowing his own role in developing COVID vaccines, perhaps his most acclaimed success and which he had called a miracle, saying of his rival for the 2024 Republican presidential nomination, "The real Ron [DeSantis, governor of Florida] is a RINO [Republican in name only] who 'Loved the Vaccines.'"[21] If their base embraces vaccine denial, Republican candidates for president must do the same or risk losing.

When the next pandemic arrives, which it will and perhaps sooner than we think, it will meet a well-established and aggressive movement of deniers who are anti-vaccination, anti-masking, anti-medicine, and anti-science. It is a fear-inducing thought that even as COVID-19 wanes, the boost it gave to the anti-vaccination movement may remain as its legacy, causing unnecessary deaths and possibly the return of diseases thought to have been virtually eradicated, such as polio, measles, mumps, rubella, and whooping cough.

The Tropical Trump

Those who deplore President Donald Trump's inept and lethal performance on COVID-19 might well imagine that no democratically elected leader could have done worse. To find a more extreme example, we might imagine, we would have to look to one of the world's dictators: Vladimir Putin or and Kim Jong Un. But that view would have been mistaken, for nearly every action of another elected president, Jair Bolsonaro of Brazil, matched and perhaps even exceeded those of Donald Trump, leading to the needless deaths of hundreds of thousands of Brazilians. Bolsonaro scoffed and joked about the danger from COVID-19, promoted quack remedies like hydroxychloroquine far longer than did President Trump, actively blocked steps that would have lowered transmission, and withheld effective treatment from Brazil's indigenous people in what some have called attempted genocide. As of March 2022, Brazil had a higher rate of deaths from COVID-19 per 100,000 than the United States, India, Russia, and every other country except Peru.[22] Moreover, the number of COVID-19 deaths in Brazil is believed to have been underreported by as much as a factor of two, whereas the United States had much lower underreporting, making the disparity even greater.[23] One group of researchers put the number of excess deaths in Brazil between January 2020 and April 2021 at 370,000.[24]

Bolsonaro's response to the COVID-19 pandemic was so corrupt, so multifaceted, that it deserves a full-length book. Indeed, in October 2021, a Brazilian congressional panel issued a nearly 1,200-page report (in Portuguese) that asserted that Bolsonaro had deliberately allowed COVID-19 to tear through the country, killing hundreds of thousands in an attempt to achieve herd immunity and restore the economy. It recommended that he be charged with "crimes against humanity" and that criminal charges be brought against his three sons and dozens of other current and former government officials.[25]

An article published in the *Journal of Public Health Policy* in August 2021 by a group of Brazilian researchers showed that hydroxychloroquine increased the length of hospital stay, the need for mechanical ventilation, and the risk of death.[26] Bolsonaro nevertheless used the Brazilian army to distribute the ineffective and dangerous drug to indigenous people. The Health Ministry also recommended ivermectin, an antiparasitic drug used to treat several rare tropical diseases. It was found to have no effect on the COVID-19 virus, as even its own manufacturer announced. The use of ivermectin may have even led to an outbreak of hepatitis.

In a speech to the nation on March 24, 2020, Bolsonaro dismissed the virus as a "little flu" and said that the "worst had passed." In a press conference two days later, he said that "Brazilians should be studied, we don't catch anything. You see people jumping in sewage, diving in it and nothing happens to them." He said that social distancing was shown to be "practically useless," and participated in anti-lockdown protests.[27] On December 17, 2020, in a speech at a political event, he joked that people taking the Pfizer–BioNTech COVID-19 vaccine might transform into crocodiles, become superhuman, turn into bearded ladies, and develop effeminate voices.[28] His administration dragged its feet for months over the purchase of vaccines, then made a deal to buy an unapproved shot from the Indian company Bharat Biotech. The Brazilian Parliamentary Inquiry Committee (CPI) opened an investigation that showed that the price initially quoted for the vaccine had risen by more than tenfold in the signed contract. The Ministry of Health then had no choice but to suspend the purchase.

Bolsonaro frequently had contact with the public without wearing a mask and encouraged others to follow his example. In one instance, he removed the mask of a ten-year-old girl posing beside him for a photograph. His sons and government officials visited indigenous communities without wearing masks.

On July 7, 2020, Bolsonaro announced that he had tested positive for the virus.[29] He tested negative a few weeks later. In September, he proclaimed that vaccination would not be mandatory and that no vaccine would be administered until it had been proven in Brazil.[30] In November, he said he would not take a vaccine and labeled masks "the last taboo to fall."[31]

Bolsonaro had been elected president of Brazil in 2018 partly in response to the Petrobras oil scandal, which had led to the conviction of former president (2003–2011) Luiz Inâcio da Silva (Lula). Lula was prohibited from running in the 2018 election, tilting the field toward Bolsonaro, who won with 55 percent of the vote. By the time of the 2022 election, a Supreme Court judge had dismissed the corruption charge against Lula, allowing him to run again. Bolsonaro repeated his long-standing criticisms of the Brazilian electronic voting system, claiming that it was open to fraud and saying that he might not recognize the result if he lost. "I have three alternatives for my future," he said: "being arrested, killed or victory."[32] He lost to Lula in a runoff, making Bolsonaro the first incumbent Brazilian president in thirty years not to win reelection. Lula was inaugurated on January 1, 2023, and by custom, the outgoing Brazilian president is expected to hand the presidential sash to his successor. But like Donald Trump, who broke a 152-year tradition by not attending the inauguration of President Joseph Biden, Bolsonaro did not show up. Facing numerous investigations, he soon fled Brazil for Orlando, Florida. In another unfortunate similarity to events in the United States, on January 8, 2023, thousands of Brazilians besieged government buildings in the capital city, Brasilia, claiming election fraud. To clear the buildings took security forces five hours. Brazil's Supreme Court quickly announced that it would investigate Bolsonaro for inspiring the riot, another similarity between the two former presidents, as Trump's role in the January 6, 2021, attack on Congress came under investigation.

13

Global Warming: The Ultimate Triumph of Science Denial?

> "The concept of global warming was created by and for the Chinese in order to make U.S. manufacturing non-competitive."[1]
>
> —Donald Trump, 2012

Manmade global warming is one of the oldest scientific theories still extant. Since the 1850s, scientists have known that carbon dioxide (CO_2) in the atmosphere preferentially absorbs heat rays rising from the earth, blanketing the planet and raising surface temperature. In the 1890s, calculations by future Nobel laureate Svante Arrhenius showed that changes in atmospheric CO_2 levels were large enough to explain the mysterious waxing and waning of the Ice Ages. Not until the late 1950s, however, were scientists able to measure the amount of CO_2 the atmosphere holds. In 1958, CO_2 amounted to 315 parts per million (by volume); today the figure stands at 420 ppm. Never once since measurements began has the amount of CO_2 in the atmosphere declined from one year to the next. As long as we keep burning fossil fuels, that rise will continue as far into the future as we can see, temperatures will soar, arctic ice will melt, sea level will rise, extreme weather will increase, some diseases will worsen, and hordes of climate migrants will be on the move. Despite this forewarning, largely because of state science denial, nations have been unable even to slow the rise of CO_2 in the atmosphere, much less end it. Global emissions from fossil fuel combustion took a slight dip in 2020, reflecting the effects of the pandemic, and then in 2021 rose to an all-time high of 37.12 billion tons.

Should humanity prove unable to contain global warming, the death toll will far exceed that of the other examples of state science denial we have reviewed put together. Some scientists believe that global warming threatens human civilization itself.

The Politicization of Global Warming

The problem we in America face in acting to curtail global warming is that, as with almost every important issue today, it has become thoroughly politicized. But this is a relatively recent development. In a February 1965 speech, President Lyndon Johnson warned that, "This generation has altered the composition of the atmosphere on a global scale through radioactive materials and a steady increase in carbon dioxide from the burning of fossil fuels."[2] Johnson's statement was based on a report by the Environmental Pollution Panel of the President's Science Advisory Committee, a group that President Eisenhower established in 1957. The 1965 report concluded:

> Through his worldwide industrial civilization, Man is unwittingly conducting a vast geophysical experiment. Within a few generations he is burning the fossil fuels that slowly accumulated in the earth over the past 500 million years. The CO_2 produced by this combustion is being injected into the atmosphere; about half of it remains there. The estimated recoverable reserves of fossil fuels are sufficient to produce nearly a 200% increase in the carbon dioxide content of the atmosphere.
>
> By the year 2000 the increase in atmospheric CO_2 [compared to the pre-industrial era] will be close to 25%. This may be sufficient to produce measurable and perhaps marked changes in climate and will almost certainly cause significant changes in the temperature... of the stratosphere.[3]

The 25 percent projection turned out to be an underestimate, as compared to a pre-industrial benchmark of about 280 ppm, by 2000 atmospheric CO_2 had risen by 32 percent. But nothing was done in response to the committee's warning, illustrating the difficulty we have in dealing with problems that lie a few decades in the future. The first action on atmospheric pollution came in April 1988, when President Ronald Reagan signed the Montreal Protocol, an international treaty aimed at reducing the depletion of ozone in the atmosphere. The harmful substances the protocol banned did not include carbon dioxide, but rather aimed at ozone-depleting refrigerants such as Freon. President Reagan said that "the protocol marks an important milestone for the future quality of the global environment and for the health and well-being of all peoples of the world."[4] The Senate passed the Montreal protocol unanimously.

In June 1988, James Hansen, director of NASA's Goddard Institute for Space Studies and an expert in computerized climate modeling, testified on a sweltering day to the U.S. Senate Committee on Energy and Natural Resources. At the time, the four warmest years on record had all occurred in the 1980s. Hansen gave the most dramatic warning that any scientist had yet offered in a public government forum, saying that he was 99 percent certain that greenhouse gases derived primarily from fossil fuels, rather than natural variation, had caused the warming trend. "It is time to stop waffling so much and say that the evidence is pretty strong that the greenhouse effect is here," Hansen said.[5] He became the villain scientist for the climate deniers, who criticized and scoffed at his projections of temperature rise. But those projections turned out to be as accurate as they could have been at the time Hansen made them.

Growing recognition of the threat from global warming led the United Nations in June 1992 to convene an Earth Summit in Rio de Janeiro, Brazil. From it emerged three "conventions," or treaties, including the United Nations Framework Convention on Climate Change. Its goal was to stabilize "greenhouse gas concentrations in the atmosphere at a level that would prevent dangerous anthropogenic interference with the climate system."[6] The convention was signed by 154 nations, including the United States, represented at the summit by President George H. W. Bush.

The first result of the Rio summit was the 1997 Kyoto Protocol, which called for quite modest, voluntary reductions in carbon emissions. It was signed by 192 countries, but the U.S. Senate unanimously passed a resolution objecting to any international climate agreement that did not require developing countries to reduce carbon emissions, as the protocol did not, on the grounds that otherwise America would be put at a competitive disadvantage. This conveniently ignored the fact that, historically, developed nations had contributed far more to global warming than less-developed ones. Knowing that a vote would fail, President Bill Clinton did not bother asking the Senate to ratify the Kyoto Protocol, and the United States never signed it. By the turn of the century, a partisan divide on global warming was on the rise and the opportunity for climate legislation was fast waning.

George W. Bush, son of the former president, was elected president by a razor-thin margin in 2000. He equivocated on global warming, telling *People Magazine* in July 2006, during his second term, "I think there is a debate about whether [global warming is] caused by mankind or whether it's caused naturally, but it's a worthy debate. It's a debate, actually, that I'm in the process of solving."[7] By that time, as scientists were approaching a unanimous consensus on global warming, climate denial was becoming a litmus test for elected Republicans. In the 2008 presidential campaign, only one Republican candidate, Senator John McCain, was willing publicly to accept manmade global warming as a serious threat. In a campaign video titled "Climate Change and the Environment," McCain said, "For years I have been bitterly disappointed in the Bush administration failures to act in an effective fashion on [global warming.]" He endorsed nuclear energy and ended the video by quoting British Prime Minister Tony Blair: "Suppose if we are wrong and there's no such thing as climate change and we adapt green technologies. All we've done is give our children a cleaner world. Suppose we are right about climate change and do nothing then what have we done for our kids."[8] McCain won the nomination but lost the election.

In 2009, at the beginning of President Barack Obama's first term, the U.S. House of Representatives passed the American Clean Energy and Security Act by the narrow margin of 219 to 212. The bill would have created a "cap-and-trade" scheme to limit the right of companies to emit specified pollutants including CO_2 but allow them to trade those rights. The European Union already had such a system in place. Senate Republicans threatened a filibuster, and the bill was never brought to the floor.

One of the opponents of cap-and-trade legislation, or any action on global warming, was an organization called Americans for Prosperity, founded in 2004 by the conservative Koch brothers, Kansas-based energy billionaires. In 2008, the organization circulated a "No Climate Tax Pledge" to federal, state, and local government officials. It read: "I (name) pledge to the taxpayers of the state of (name) and to the American people that I will oppose any legislation relating to climate change that includes a net increase in government revenue."[9] The last phrase would apply to cap-and-trade, to a carbon tax, or to any similar measure. Within two years, 165 members of Congress and candidates had signed the pledge and by 2013, the number had grown to 411 politicians nationwide, all Republicans except for 4 Democrats and 2 Independents. Global warming denial had become a signature issue—literally—for the Republican party. One effect has been to make outright denial of global warming by Republican members of Congress no longer necessary. With the entire party in lockstep on the issue, members for the most part simply ignore it as beneath discussion and there was little the Democrats could do about it.

This left for President Obama only the route of executive action. On September 3, 2016, he signed the Paris Agreement on climate, as did President Xi of China. "Someday," Obama said, "We may see this as the moment that we finally decided to save our planet."[10] The agreement had been negotiated by 196 countries at the 2015 United Nations Climate Change Conference held near Paris. Its goal was to limit the rise in global temperature to 1.5°C above the pre-industrial level, which it was hoped would avoid the worst effects of global warming. But to achieve that goal would require cutting global emissions by about 50 percent by 2030 and by 100 percent by 2050. Some reduction pathways could meet that schedule, but by the early 2020s, many scientists had come to believe that holding the temperature rise to 1.5°C is no longer attainable, while a 2°C limit might still be possible if action were to come soon. Then in 2016 came the election of Donald Trump, who in 2012 had tweeted that "the concept of global warming was created by and for the Chinese in order to make U.S. manufacturing non-competitive."[11] The online magazine *Mother Jones* has tracked Trump's dozens of statements related to global warming from December 2009 to December 2018.[12] During this period, he took every position imaginable, but his default was clearly that global warming is false. Half of the appointees to his twenty-four-person cabinet were on record as denying manmade global warming. Some, like Environmental Protection Agency Head Scott Pruitt, had been among the most active climate deniers, in his case repeatedly suing the agency he would lead—and decimate.

In June 2017, President Trump withdrew the United States from the Paris agreement, placing the country among the handful of nations that had not signed: Eritrea, Iran, Iraq, Libya, and Yemen.[13] His announcement of the decision focused entirely on the alleged damage that adherence to the agreement would do the U.S. economy, with no sense of America's role in the community of nations or to the long-term threat. Nowhere did Trump mention global warming, the problem the Paris Agreement was designed to address. Nearly every assertion in his statement was at best a distortion, and at worst an outright lie, yet many listeners would have had no way of knowing that and were likely persuaded that withdrawal was the right thing to do.

At the end of his statement, Trump called on Pruitt to say a few words. He applauded Trump's action, saying, "Thank you, Mr. President. Your decision today to exit the Paris Accord reflects your unflinching commitment to put America first. Mr. President, it takes courage, it takes commitment to say no to the plaudits of men while doing what's right by the American people." Pruitt's fawning hero worship did not quite reach the heights of the Soviet speaker who had applauded Stalin as "the first agronomist of the whole world," but the sentiment was the same.[14] Pruitt would resign in disgrace on July 5, 2018, under a cloud of ethics investigations. By that time, he had gutted the EPA, weakened its advisory committees, and filled its upper ranks with science deniers, reminiscent of Lysenko's actions at the All-Union Lenin Academy of Agriculture. Even as this book is being written, the EPA has still not recovered from the departure of so many scientists and policy experts.

By the beginning of Trump's term of office, he and virtually all Republican officeholders had decided as a matter of ideology that manmade global warming is false. A signal moment that clearly captured this stance came in Trump's January 2018 State of the Union message, when he turned to energy policy and said, "We have ended the war on American energy—and we have ended the war on beautiful, clean coal. We are now very proudly an exporter of energy to the world."[15] The entire Republican side leapt to their feet in applause, and smiles broke out on the faces of the Cabinet members in attendance. But the statement was misleading at best, as President Obama had not conducted a "war on coal" and the United States was not a net energy exporter. But the words did not matter, only what they symbolized: the triumph of climate denial as U.S. state policy. Again, we are reminded of Stalin's Soviet Union and the standing ovation that Lysenko received at the 1948 conference when he announced that the Politburo had banned genetics.

The required delay period for leaving the Paris Agreement meant that the U.S. withdrawal could not take effect until 2020. On his first day in office, January 20, 2021, President Joe Biden signed an executive order re-admitting the United States. On August 6, 2022, Democrats in Congress introduced the Inflation Reduction Act (IRA), which included major spending to combat global warming. It passed the House 220 to 207, with every Democrat voting yes and every Republican voting no. The measure had appeared to be in trouble in the Senate, where the 50-50 tie between Republicans and Democrats meant that passage would allow no Democrat defections and would also require the tie-breaking vote of Vice President Kamala Harris. Such unanimity appeared unlikely, as West Virginia Democrat Senator Joe Manchin, who made his fortune selling coal mixed with rock and clay but that can still be burned to generate electricity, had threatened to block any climate bill. Then to nearly everyone's surprise, Manchin reversed himself, allowing the IRA to pass the Senate 51 to 50, with every Republican again voting no. The bill was the largest piece of legislation ever to address climate change in any country—a true landmark. The investment bank Credit Suisse estimates total federal climate spending under the IRA at $800 billion.[16] The act includes funding for renewable energy including nuclear power, improved grid storage of electricity, home energy efficiency upgrades, incentivizing the purchase of electric vehicles, and much more. The bill also restored the authority of the EPA to regulate CO_2 and other greenhouse gases, which the Supreme Court had prohibited on June 30, 2022, in a 6-3 vote.

Had the IRA failed to pass, who could say when another bill addressing global warming might win a majority? Not in the 118th Congress that began on January 3, 2023, for no climate-related bill could get past the narrow Republican majority in the House. But pass the IRA did, providing breathing space for the future action on global warming that will surely be necessary.

How can we explain how one political party could vote unanimously for the IRA, with its climate provisions, and the other vote unanimously against it? The answer has several components:

- Republicans have long been ideologically opposed to government regulation of industry, which action on global warming would almost certainly entail. One of President Reagan's best-known quotes was, "The most terrifying words in the English language are: I'm from the government and I'm here to help." He and other Republicans believed that too much government regulation stifles entrepreneurship and innovation and that the less government interference in business the better. Let the market work its magic.
- Regulation of carbon emissions would target not some minor and easily replaced segment of U.S. industry, but rather the giant energy companies that depended on fossil fuels. Yet those very fuels had been largely responsible for the rise of the modern industrial state.
- Until this century, it appeared unlikely that renewable sources of energy could replace fossil fuels in the foreseeable future. To cut back on fossil fuel consumption would then have meant an overall decline in energy usage. That in turn would damage the economy and require unwelcome changes in our way of life. (Note that as the cost of renewables has fallen, these outcomes have been largely avoided. Since 2010, the cost of building large wind farms has dropped more than 40 percent, and the cost of solar energy has declined by more than 80 percent. These trends have allowed Britain, for example, to go for long periods without using coal power. The United States will likely not be far behind.[17])
- The closer the margins in Congress, the less each party can afford defections. This leads to a rigid party line that no member

can cross without risking his or her standing. Thus even those Republicans who might wonder whether scientists might not be right about global warming had to swallow their doubts.
- These factors dovetailed with another that may have been even more important: the claim that manmade global warming is false, or at best unproven, in which case it makes sense to wait for so-called sound science. This despite a near-unanimous consensus among scientists that manmade global warming is true. Climate deniers were led to this view by one of the world's largest corporations.

Merchants of Doubt[18]

Hard as it is to believe today, one of the first to confirm the reality and danger of manmade global warming was the Exxon Corporation.[19] In July 1977, Exxon senior scientist James Black told the company's executive managers, "There is general scientific agreement that the most likely manner in which mankind is influencing the global climate is through carbon dioxide release from the burning of fossil fuels."[20] The next year he reported that doubling the concentration of atmospheric CO_2 would raise average global temperature by two or three degrees, close to the projection from today's supercomputer climate models (and from Arrhenius's 1890s calculation.) Instead of acting responsibly to this warning, Exxon adopted the strategy of doubt and delay that Big Tobacco used successfully in its forty-year denial of a link between smoking and lung cancer.[21] In 1989, ExxonMobil (renamed after a merger of the two companies) founded the Global Climate Coalition, made up of businesses opposed to regulation of greenhouse gas emissions. One of the most damning documents of this period came from the American Petroleum Institute, the industry's trade association and a member of the coalition. In a 1998 memo, the institute "mapped out a multifaceted deception strategy for the fossil fuel industry... outlining plans to reach the media, the public, and policy makers with a message emphasizing 'uncertainties' in climate science."[22] According to the memo, "victory" would be achieved for the campaign when "average citizens" and the media became convinced of "uncertainties" in climate science.[23]

These early documents showed that some Exxon scientists were as aware of global warming as those in academe and government. But no one outside the company had ever dug deep enough to discover "what exactly did oil and gas companies know, and how accurate did their knowledge prove to be?" That was the purpose of an article published in early 2023 with the title "Assessing ExxonMobil's global warming projections."[24] As shown in Figure 13.1, taken from the article, ExxonMobil's predictions were right on the money.

Instead of making this important information public, ExxonMobil executives hid it and lied about it, "overemphasizing uncertainties, denigrating climate models, mythologizing global cooling, feigning ignorance about the discernibility of human-caused warming, and staying silent about the possibility of stranded fossil fuel assets in a carbon-constrained world."[25] ExxonMobil's executives repeatedly

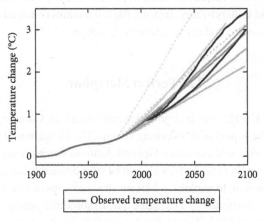

Figure 13.1 Temperature projections by Exxon scientists between 1977 and 2003 (black and gray) compared with the observed rise shown in red, which reached 1.17°C in 2020. Note that each of Exxon's projections shows temperature increasing by more than 2°C by 2100. Some are even higher than today's projections by the Intergovernmental Panel on Climate Change.[a]

[a] Supran, "Assessing ExxonMobil's Global Warming Projections."

misled the public, their own investors, and Congress. As late as 2013, twenty-five years after James Hansen's testimony and after a mountain of evidence had confirmed manmade global warming, ExxonMobil CEO Rex Tillerson, who would become Trump's secretary of state, said in a television interview: "The facts remain there are uncertainties around the climate... what the principal drivers of climate change are ... there are other elements of the climate system that may obviate this one single variable [of burning fossil fuels].... And so that's that uncertainty issue."[26]

To spread their message of denial and delay, since 1988 ExxonMobil has awarded $33 million to right-wing "think tanks" and policy associations, topped by the American Enterprise Institute at $4,199,000.[27] Today, a number of lawsuits, political investigations, and campaigns are underway in an effort to hold ExxonMobil and other fossil fuel corporations accountable for misleading the public and their own shareholders about the dangers of global warming. These companies could have begun decades ago to transition to renewable energy sources, including nuclear power, which could have made them world leaders by now. Instead, Big Oil remains committed to find more oil reserves and to extracting the last drop.

The Perfect Metaphor

On April 4, 2022, the Intergovernmental Panel on Climate Change released the report of its Working Group III, "Mitigation of Climate Change," with UN Secretary-General António Guterres saying that "investments [in fossil fuels] will soon be stranded assets [reserves that are uneconomic to exploit], a blot on the landscape, and a blight on investment portfolios."[28] That same day, ExxonMobil announced that it would spend $10 billion to develop a new oil and gas project in the South American country of Guyana.

On March 19, 2023, the IPCC published its sixth assessment Synthesis Report, which integrates the main findings of three working groups and other special reports. We can think of this report as the culmination of thirty-five years of work by thousands of scientists since the founding of the IPCC in 1988. Throughout its history, the

IPCC has proceeded cautiously, trying to avoid accusations of being "alarmist." But this has led many scientists to believe that it has understated the threat. But no one could call the following statement from the 2023 Synthesis overly cautious or fail to understand its import: "To limit warming to 1.5°C ... [or] to 2°C, involve[s] rapid and deep and, in most cases, immediate greenhouse gas emissions reductions in all sectors *this decade* (emphasis added)."[29] Scientists can speak no more clearly than that.

At a New Hampshire town hall meeting on February 29, 2020, in response to a question about oil drilling in Alaska, candidate Joe Biden pledged, "No more drilling on federal lands, period, period, period." Drilling for oil is Alaska, he added, would be "a disaster, a big disaster, in my view."[30] During the Democrat candidates' televised debate on March 15, 2020, Biden promised that if he were elected, there would be "no ability for the oil industry to continue to drill, period."[31]

Yet three years later, on March 14, 2023, the Biden administration gave approval to ConocoPhillips for the giant Willow oil development project in the U.S. National Petroleum Reserve (NPRA), on the frozen tundra of Alaska's north coast, which had been set aside by the government in 1923 to ensure a domestic supply of oil. This action put Biden ahead of Donald Trump in the number of drilling leases each permitted during their first twenty-five months in office. ConocoPhillips expects to extract 600 million barrels of oil over the next three decades. As the *Guardian* points out, this amounts to adding to the atmosphere the emissions of an entire country, such as Belgium.[32] Or as *Politico* observed, the project is "the equivalent of adding two new coal-fired power plants to the U.S. electricity system every year."[33] The reason for the broken promises, according to *Politico*, was to fashion a "Biden-moves-to-the-center" narrative to improve his chances in the 2024 presidential election.[34]

In addition to the bare fact of the 300 million tons of CO_2 to be emitted over the lifetime of the Willow Project, the symbolism must dishearten anyone who took heart at Biden's election, his return of the United States to the Paris Agreement, and the passage of the IRA. One such was former Vice President Al Gore, who said that approving the Willow Project was "recklessly irresponsible" and "a recipe for climate chaos."[35] The Willow Project made news, but it amounts to only one-third of the total amount of oil and gas the United States has on its books to produce. That figure excludes shale fracking, but if it is included, America plans to extract more oil and gas than any other nation.[36]

Since the beginning of the twentieth century, temperature in the Arctic has risen three times faster than the global average. A project that will last for thirty years on the North Slope of Alaska thus must take global warming into account in its planning, especially the near certainty of melting permafrost. To protect its heavy equipment from sinking into the soft, thawing ground, ConocoPhillips plans to install chillers to keep the permafrost frozen. One hardly knows whether to laugh or cry at this news. As author Elizabeth Kolbert put it, "A massive oil project that requires chilling the permafrost is, unfortunately, the perfect metaphor for our time."[37] Until we and those we elect can gather the collective will to say no to Big Oil, we have little hope of limiting global warming. Then science denial will have achieved its ultimate triumph, at the cost of many if not most of the benefits science has brought humanity, possibly even human civilization itself.

14
Roadmap to Doomsday

State-level science denial can occur in two ways: top-down, by order of dictators and by the ignorance and perceived self-interest of elected officials, and bottom-up, driven by mass movements of people and now amplified by the echo-chamber effect of the internet. However, they share one thing in common: they are based on fallacies. Here are some examples we have encountered of conclusions stemming from such fallacies: acquired traits are heritable, genes do not exist, cuckoos can hatch from the eggs of warblers, wheat can grow thick enough to support the weight of children, anti-retroviral drugs cause AIDS and do not cure it, Jewish scientists are genetically inferior, euthanasia can perfect a master race, COVID-19 is a "little flu," vaccination is not a cure for disease but a cause, and global warming is a Chinese hoax. Most of these are obviously false on their face, while others wither under a modicum of common sense. It is shocking to realize that tens of millions of people have died unnecessarily simply because of such lies.

Science is fundamentally a search for the truth about the natural world. Thus science and lies are profoundly incompatible. Of course, scientists are human. We make mistakes, and some have even falsified their data. But science has built-in lie detectors. One is peer review: To appear in a scientific journal, an article must be scrutinized and passed by editors and by experts (the peers). They look to see whether the evidence backs up the claims made in an article. The other safety feature of science is replication. Scientific articles must provide enough information about the methods used so that someone else could replicate the findings. A classic case was the 1989 claim by electrochemists Martin Fleischmann and Stanley Pons that they had observed a nuclear reaction at room temperature, known as cold fusion. If true, they would have discovered the holy grail of cheap, clean, sustainable energy. One clue that the claim might not hold up was that it was announced at a press conference rather than in a peer-reviewed scientific journal. Another was that they provided little information about the research methods they had used. And sure enough, other scientists soon reported that they could not produce cold fusion in a laboratory setting.

Of course, truth is essential in all human affairs. As individuals, the way we think, make decisions, and act are all influenced by our understanding of what we believe to be true. If we blind ourselves to the truth, we make poor decisions and thus cause suffering for ourselves, our loved ones, and for society as a whole. If we cannot believe what those close to us say, then marriage and friendship, the underpinnings of human society, break down. We all know this.

In the broad societal context, truth is just as fundamental. The values that sustain civil societies, such as fairness, equality, and justice, depend on the upholding of truth. The most obvious example may be courts of law, where witnesses swear to tell the truth, the whole truth, and nothing but the truth. Without that, miscarriages of justice would be the rule and civil society could not function. Democracy also depends on truth. When people have accurate information, they can make informed decisions about which candidates to vote for. They can hold their elected leaders accountable for the promises they made. When voters are uncertain about the truthfulness of political candidates, they may base their decisions on how closely candidates align with their own ideology and beliefs. When ideology replaces truth, as is happening widely today, partisan conflict becomes the rule, leading to gridlock and the failure of democracy.

On the larger scale of international diplomacy and cooperation, truth is essential to comprehend the dynamics of a globalized world, to negotiate treaties, and to avoid misunderstandings that can lead to war. It is striking that most of the recent conflicts have been justified by falsehoods. Here's the truth: Saddam Hussein did not have weapons of mass destruction; Iraq was not responsible for the World Trade Center disaster; Ukraine has not always been a part of Russia and today is rightfully a distinct nation with its own language, culture, and aspirations for independence.

We distinguish truth from falsehood by using reason and our common sense, weighing evidence and reaching logical conclusions.

In science, one of the most critical uses of reason, if not its most important, is to observe effects and to develop and test theories to identify their cause. This method has proven itself when applied on the gargantuan scale of galaxies down to the subatomic dimensions of the evanescent Higgs boson.

In the current era, telling truth from falsehood has become increasingly difficult. The internet offers an immense volume of readily accessible material, but with no way to tell what is true and what is not. Whatever your position on any issue, whether right or wrong, you are likely to find it reinforced on the internet by others with whom you can band together electronically. If you prefer to get your information from television, you can likely find a news network that agrees with and reinforces your political stance regardless of the truth.

We began this book by noting how science, along with its partners engineering and medicine, has extended human life and enhanced its quality. Despite this success, during the twentieth century science denial cost scores of millions of lives. Already in the twenty-first century, when we might have hoped that science denial would wither away, it has grown even stronger. The rejection of medical science during the COVID-19 pandemic led to the unnecessary loss of hundreds of thousands of lives. Moreover, it undercut confidence in medicine generally, with unknowable but surely harmful long-term consequences. Over the first two decades of the century, scientists have nearly unanimously concluded that manmade global warming is true. The evidence has grown to be overwhelming, and there is not a shred of scientific evidence against manmade global warming. What has been the Republicans' response? To make denial of global warming synonymous with their party. Nothing makes this clearer than a shocking set of policy recommendations from the conservative Heritage Foundation in its Project 2025, which lays out an energy agenda for the first 180 days of the next Republican administration.[1] These include:

- Eliminate regulations aimed at reducing greenhouse gas emissions from cars, oil and gas wells, and power plants.
- Dismantle nearly all federal clean energy programs.

- Repeal the Inflation Reduction Act of 2022 and its breakthrough funding to combat global warming, including support for electric vehicles.
- Shutter the Department of Energy office that funds loans for green technologies.
- Remove climate change from the agenda of the National Security Council.
- Encourage allied nations to buy and use more fossil fuels rather than renewable energy.

For humanity, this is nothing less than a roadmap to doomsday. For the sake of our children and grandchildren, we must reject it and vote only for politicians who accept and prioritize science.

Acknowledgments

With deepest thanks to my agent of long-standing, John Thornton, to Jonathan Cobb for expert editing, and to Carol Shetler for proofreading.

Notes

Epigraph
1. Timothy Ferris, "The Space Telescope: A Sign of Intelligent Life," *The New York Times*, April 29, 1990.

Chapter 1
1. WHO, "A Brief History of Vaccination," https://www.who.int/news-room/spotlight/history-of-vaccination/a-brief-history-of-vaccination
2. T. Hobbes and H. Morley, *Leviathan: Or, The Matter, Form, and Power of a Commonwealth Ecclesiastical and Civil*, Morley's Universal Library (G. Routledge and Sons, 1887), https://books.google.com/books?id=-l6x3KZt-0cC
3. Science denialism, a term sometimes used instead of denial, goes beyond the individual rejection of scientific knowledge and refers to organized efforts to undermine or discredit a scientific consensus.
4. I do not include biblical creationism in this book because I am focusing on science denial that has cost large numbers of lives.

Chapter 2
1. D. Joravsky, *The Lysenko Affair* (University of Chicago Press, 1986), https://books.google.com/books?id=K7j-mAEACAAJ, 58–59.
2. Valery N. Soyfer, "The Consequences of Political Dictatorship for Russian Science," *Nature Reviews Genetics* 2, no. 9 (2001): 723–729.
3. Soyfer, "The Consequences of Political Dictatorship for Russian Science," 727.
4. Horsley Gannt, "Ivan Pavlov," in *Encyclopedia Britannica*, n.d., https://www.britannica.com/biography/Ivan-Pavlov
5. Richard Cavendish, "Death of Ivan Pavlov | History Today," 2011, https://www.historytoday.com/archive/death-ivan-pavlov
6. Soyfer, "The Consequences of Political Dictatorship for Russian Science," 728.
7. Joravsky, *The Lysenko Affair*, 58–59.
8. Marc Somssich, "A Short History of Vernalization," February 9, 2020. https://doi.org/10.5281/ZENODO.3660691, 10.
9. S. J. Gould, *The Structure of Evolutionary Theory* (Harvard University Press, 2002), https://books.google.com/books?id=lLkFAwAAQBAJ, 62.
10. J. S. Huxley, M. Pigliucci, and G. B. Muller, *Evolution, the Definitive Edition: The Modern Synthesis* (MIT Press, 2009), https://books.google.com/books?id=FxtFAQAAIAAJ
11. "Trofim Lysenko | Soviet Biologist and Agronomist | Britannica," https://www.britannica.com/biography/Trofim-Lysenko
12. V. Soyfer, *Lysenko and the Tragedy of Soviet Science* (New Brunswick, NJ: Rutgers University Press, 1994), 11.
13. P. Pringle, *The Murder of Nikolai Vavilov: The Story of Stalin's Persecution of One of the Great Scientists of the Twentieth Century* (New York: Simon & Schuster, 2008), https://books.google.com/books?id=2mi7NAOxe-8C

14. Joravsky, *The Lysenko Affair*, 2010, 60.
15. Joravsky, 1986, 191.
16. N. Roll-Hansen, *The Lysenko Effect: The Politics of Science*, Control of Nature (Amherst, NY: Humanity Books, 2005), https://books.google.com/books?id=TP4PAQAAMAAJ, 68.
17. Joravsky, *The Lysenko Affair*, 1986, 60.
18. Soyfer, "The Consequences of Political Dictatorship for Russian Science," 19.
19. Joravsky, *The Lysenko Affair*, 2010, 62.
20. Joravsky, 1986, 85–86.
21. Pringle, *The Murder of Nikolai Vavilov*, 150.
22. Pringle, 150.
23. Joravsky, *The Lysenko Affair*, 1986, 37.
24. Pringle, *The Murder of Nikolai Vavilov*, 151.
25. Joravsky, *The Lysenko Affair*, 2010, 37.
26. Joravsky, 1986, 37.

Chapter 3

1. Roll-Hansen, *The Lysenko Effect*, 97.
2. Pringle, *The Murder of Nikolai Vavilov*, 154.
3. Pringle.
4. Pringle.
5. Joravsky, *The Lysenko Affair*, 374.
6. Soyfer, *Lysenko and the Tragedy of Soviet Science*, 54.
7. Soyfer, 193.
8. J. Becker, *Hungry Ghosts: Mao's Secret Famine*, A Holt Paperback: History (New York, NY: Henry Holt and Company, 1998), https://books.google.com/books?id=3VeLKJyRzuQC, 57.
9. Steven Bela Vardy and Agnes Huszar Vardy, "Cannibalism in Stalin's Russia and Mao's China," *East European Quarterly* 41, no. 2 (2007): 223.
10. Amartya Sen, *Poverty and Famines: An Essay on Entitlement and Deprivation* (Oxford University Press, 1982), 1.
11. Pringle, *The Murder of Nikolai Vavilov*, 191
12. Pringle, 200.
13. Pringle, 201.
14. Roll-Hansen, *The Lysenko Effect: The Politics of Science*, 97.
15. Pringle, *The Murder of Nikolai Vavilov*, 198.
16. Soyfer, *Lysenko and the Tragedy of Soviet Science*, 69.
17. Joravsky, *The Lysenko Affair*, 40.
18. Joravsky, 43.
19. Soyfer, *Lysenko and the Tragedy of Soviet Science*, 64.
20. Soyfer, 70.
21. Soyfer, 73.
22. Soyfer, 81.
23. Soyfer, 82.
24. Soyfer, 81–83.
25. Zhores A. Medvedev, *The Rise and Fall of T. D. Lysenko* (New York: Columbia University Press, 1969), 25.
26. Soyfer, *Lysenko and the Tragedy of Soviet Science*, 81–87.
27. Pringle, *The Murder of Nikolai Vavilov*, 211. The meeting was postponed to 1939 and took place in Edinburgh. Vavilov was elected president in absentia, his seat left vacant to make his absence unmistakable.

Chapter 4

1. Soyfer, *Lysenko and the Tragedy of Soviet Science*, 136.
2. Soyfer, 76.
3. Soyfer, 75–76.
4. Sam Kean, "The Soviet Era's Deadliest Scientist Is Regaining Popularity in Russia," *The Atlantic*, December 19, 2017, https://www.theatlantic.com/science/archive/2017/12/trofim-lysenko-soviet-union-russia/548786/
5. Soyfer, "The Consequences of Political Dictatorship for Russian Science," 80.
6. Nathan Nunn and Nancy Qian, "The Potato's Contribution to Population and Urbanization: Evidence from an Historical Experiment" (Cambridge, MA: National Bureau of Economic Research, July 2009), https://doi.org/10.3386/w15157
7. Joravsky, *The Lysenko Affair*, 1986, 280.
8. Joravsky, 277.
9. Joravsky, 279.
10. N. S. Khrushchev and E. Crankshaw, *Khrushchev Remembers* (Boston: Little, Brown, 1970), https://archive.org/details/khrushchevrememb0000edwa, 99–100.
11. R. Conquest, *The Great Terror: A Reassessment* (Oxford University Press, 1990), https://books.google.com/books?id = 16l79hfKMzEC, 359.
12. Joravsky, *The Lysenko Affair*, 1986, 317.
13. Soyfer, *Lysenko and the Tragedy of Soviet Science*, 120.
14. Salil Sen, "Speech Delivered at a Reception in the Kremlin to Higher Educational Workers," 2008, https://www.marxists.org/reference/archive/stalin/works/1938/05/17.htm
15. Soyfer, *Lysenko and the Tragedy of Soviet Science*, 129–130.
16. Soyfer, 130.
17. Roll-Hansen, *The Lysenko Effect*, 253.
18. Soyfer, *Lysenko and the Tragedy of Soviet Science*, 129.
19. Medvedev, *The Rise and Fall of T. D. Lysenko*, 58.
20. Soyfer, *Lysenko and the Tragedy of Soviet Science*, 127.
21. Medvedev, *The Rise and Fall of T. D. Lysenko*, 60–63.
22. Medvedev, 63.
23. Soyfer, *Lysenko and the Tragedy of Soviet Science*, 132–135.
24. Soyfer, 134–135.
25. Soyfer, 135.
26. Soyfer, 99.
27. Soyfer, 100.
28. Soyfer, 100.
29. Pringle, *The Murder of Nikolai Vavilov*, https://books.google.com/books?id = 2mi7NAOxe-8C, 254–258.

Chapter 5

1. Pringle, *The Murder of Nikolai Vavilov*, 2008, 290.
2. Soyfer, *Lysenko and the Tragedy of Soviet Science*, 161.
3. Soyfer, 162.
4. Soyfer, 210.
5. Medvedev, *The Rise and Fall of T. D. Lysenko*, 105.
6. Borinskaya, Ermolaev, and Kolchinsky, "Lysenkoism against Genetics: The Meeting of the Lenin All-Union Academy of Agricultural Sciences of August 1948, Its Background, Causes, and Aftermath," *Genetics* 212, no. 1 (2019): 1–12.
7. Bengt O. Bengtsson and Anna Tunlid, "The 1948 International Congress of Genetics in Sweden: People and Politics," *Genetics* 185, no. 3 (2010): 709–715.

8. Borinskaya "Lysenkoism against Genetics," 4.
9. Borinskaya, 4–5.
10. Soyfer, "The Consequences of Political Dictatorship for Russian Science," 725.
11. Kiril Rossianov, "Editing Nature: Joseph Stalin and the 'New' Soviet Biology," *Isis* 84, no. 4 (December 1993): 728–745, https://doi.org/10.1086/356638
12. S. J. Gould, *Hen's Teeth and Horse's Toes: Further Reflections in Natural History* (New York: W. W. Norton, 2010), https://books.google.com/books?id=o6g2tvN0nJoC
13. Pringle, *The Murder of Nikolai Vavilov*, 2008, 290.
14. Borinskaya.
15. Borinskaya.
16. Borinskaya, 6.
17. Joravsky, *The Lysenko Affair*, 1986, 140.
18. Borinskaya, "Lysenkoism against Genetics," 7.
19. Soyfer, *Lysenko and the Tragedy of Soviet Science*, 206.
20. Joravsky, *The Lysenko Affair*, 1986, 141.
21. Soyfer, *Lysenko and the Tragedy of Soviet Science*, 208–209.
22. N. S. Khrushchev and S. Khrushchev, *Memoirs of Nikita Khrushchev*, vol. 2 (Pennsylvania State University, 2004), https://books.google.com/books?id=uv1zv4FZhFUC, 212.
23. Soyfer, *Lysenko and the Tragedy of Soviet Science*, 254.
24. Joravsky, *The Lysenko Affair*, 1986, 184.
25. Jay Bergman, *Meeting the Demands of Reason: The Life and Thought of Andrei Sakharov* (Ithaca, NY: Cornell University Press, 2011), 109.
26. L. Graham, *Lysenko's Ghost: Epigenetics and Russia* (Harvard University Press, 2016), https://books.google.com/books?id=FTsADAAAQBAJ
27. Graham, *Lysenko's Ghost*, p. 70.
28. Graham, 74.
29. T. Snyder, *Bloodlands: Europe between Hitler and Stalin* (New York: Basic Books, 2012), https://books.google.com/books?id=maEfAQAAQBAJ
30. Snyder, *Bloodlands*, 53.
31. Snyder, 53.
32. Soyfer, *Lysenko and the Tragedy of Soviet Science*, xxiv.

Chapter 6

1. J. Becker, *Hungry Ghosts*, 54.
2. W. H. Mallory, *China: Land of Famine*, American Geographical Society. Special Publication No. 6, Ed. by G. M. Wrigley (American Geographical Society, 1926), https://books.google.com/books?id = TVuwAAAAIAAJ, 1.
3. D. C. Twitchett and J. K. Fairbank, *The Cambridge History of China*, The Cambridge History of China, vol. 11 (Cambridge University Press, 1978), https://books.google.com/books?id = 4HqP6DKE1HwC, 55.
4. "Revisiting Stalin's and Mao's Motivations in the Korean War | Wilson Center," https://www.wilsoncenter.org/blog-post/revisiting-stalins-and-maos-motivations-korean-war
5. Laurence Schneider, "Michurinist Biology in the People's Republic of China, 1948–1956," *Journal of the History of Biology* 45, no. 3 (2012): 525–556.
6. Rudolf Hagemann, "How Did East German Genetics Avoid Lysenkoism?," *Trends in Genetics* 18, no. 6 (June 2002): 320–324, https://doi.org/10.1016/S0168-9525(02)02677-X
7. Schneider, "Michurinist Biology in the People's Republic of China, 1948–1956."
8. Schneider.

9. Schneider, 530.
10. Schneider, 507–538.
11. Li Peishan, "Genetics in China: The Qingdao Symposium of 1956," *Isis* 79, no. 2 (June 1988): 227–236, https://doi.org/10.1086/354697
12. Peishan, 228.
13. Schneider, "Michurinist Biology in the People's Republic of China, 1948–1956."
14. Hagemann, "How Did East German Genetics Avoid Lysenkoism?"
15. Hagemann, 320.
16. Peishan, "Genetics in China."
17. Peishan, 233.
18. Soyfer, *Lysenko and the Tragedy of Soviet Science*, 174.
19. Khrushchev and Crankshaw, *Khrushchev Remembers*, 13.
20. Becker, *Hungry Ghosts*, 93.
21. Becker, 91.
22. Becker, 97.
23. Soyfer, *Lysenko and the Tragedy of Soviet Science*, 245.
24. Becker, *Hungry Ghosts*, 100.
25. Becker, 107.
26. Jisheng Yang, *Tombstone: The Great Chinese Famine, 1958–1962* (New York: Farrar, Straus and Giroux, 2012), https://books.google.com/books?id=bvs9cS8lzhEC
27. Yang, *Tombstone*, 3.
28. Yang, 6.
29. Yang, 13.
30. Várdy, "Cannibalism in Stalin's Russia and Mao's China."
31. Becker, *Hungry Ghosts*, 324.
32. Yang, *Tombstone*, 430.
33. F. Dikötter, *Mao's Great Famine: The History of China's Most Devastating Catastrophe, 1958–1962* (New York and London: Bloomsbury Publishing, 2010), https://books.google.com/books?id=5NsMWCHDStQC

Chapter 7

1. A. Beyerchen, *Scientists Under Hitler: Politics and the Physics Community in the Third Reich* (New Haven, CT: Yale University Press, 1977), https://books.google.com/books?id=pwoVHAAACAAJ, 23.
2. Beyerchen, *Scientists Under Hitler*, 23.
3. Beyerchen, 23.
4. Beyerchen, 25.
5. D. Stoltzenberg, *Fritz Haber: Chemist, Nobel Laureate, German, Jew* (Chemical Heritage Press, 2004), https://books.google.com/books?id=0ekNIaJX3-YC, 278.
6. Stoltzenberg, *Fritz Haber*, 280.
7. Beyerchen, *Scientists Under Hitler*, 66.
8. Beyerchen, 71.
9. Beyerchen, 71.
10. "*Deutsche Physik*," in *Wikipedia*, June 21, 2022, https://en.wikipedia.org/w/index.php?title=Deutsche_Physik&oldid=1094272798
11. Beyerchen, *Scientists Under Hitler*, 137.
12. Beyerchen, 11.
13. Beyerchen, 122.
14. "Paul Weyland," http://www.physik.uni-halle.de/Fachgruppen/history/weyland.htm
15. Beyerchen, *Scientists Under Hitler*, 130–132.

16. Philipp Lenard, *Great Men of Science*, 1933, http://archive.org/details/in.ernet.dli.2015.203547
17. Lenard, *Great Men of Science*, 37.
18. A. Beyerchen, *Scientists Under Hitler*, 175.
19. Philip Ball, "Astronomers Unknowingly Dedicated Moon Craters to Nazis. Will the next Historical Reckoning Be at Cosmic Level?" *Prospect Magazine*, June 26, 2020, https://www.prospectmagazine.co.uk/science-and-technology/astronomists-unknowingly-dedicated-moon-craters-to-nazis-will-the-next-historical-reckoning-be-at-cosmic-level
20. Beyerchen, *Scientists Under Hitler*, 219-220.
21. Beyerchen, 226.
22. Beyerchen, 230.

Chapter 8

1. Richard Weikart, "The Role of Darwinism in Nazi Racial Thought," *German Studies Review* 36, no. 3 (2013): 537–556, https://doi.org/10.1353/gsr.2013.0106, 541.
2. "Hadamar," https://encyclopedia.ushmm.org/content/en/gallery/hadamar
3. "Nazi Persecution of the Mentally & Physically Disabled," https://www.jewishvirtuallibrary.org/nazi-persecution-of-the-mentally-and-physically-disabled
4. The American government prepared a documentary film of the concentration camps that was shown as evidence at the Nuremberg trials. George Stevens directed it, and it is available on Wikimedia Commons at https://commons.wikimedia.org/wiki/File:Nazi_Concentration_Camps.webm. It is titled "Concentration Camps in Germany, 1939–1945." The segment on Hadamar begins at 14.00 minutes. Be warned: it is almost impossible to watch this film without becoming sick at your stomach.
5. Snyder, *Bloodlands*, 162.
6. C. Darwin, *The Descent of Man,: And Selection in Relation to Sex* (London: John Murray, Albemarle Street, 1871), https://books.google.com/books?id=j-8hm8eHgUC, 168–169.
7. Weikart, "The Role of Darwinism in Nazi Racial Thought," 541.
8. Editors of Encyclopaedia Britannica, "Mein Kampf" (Encyclopedia Britannica, March 2023), https://www.britannica.com/topic/Mein-Kampf
9. Luther Burbank. *The Training of the Human Plant* (New York: Century, 1907). http://archive.org/details/trainingofhumanp00burbuoft
10. Linda Villarosa, "The Long Shadow of Eugenics in America," *The New York Times*, June 8, 2022, sec. Magazine, https://www.nytimes.com/2022/06/08/magazine/eugenics-movement-america.html
11. "Indiana Eugenics: History and Legacy," https://eugenics.iupui.edu/
12. "Buck v. Bell, 274 U.S. 200 (1927)," Justia Law, https://supreme.justia.com/cases/federal/us/274/200/
13. E.S. Gosney, and P. Popenoe. *Sterilization for Human Betterment: A Summary of Results of 6,000 Operations in California, 1909–1929* (New York, London: Macmillan, 1931), 135. https://books.google.com/books?id=l_NJAAAAYAAJ
14. Corey G. Johnson, "Female Inmates Sterilized in California Prisons without Approval," Reveal, July 7, 2013, http://revealnews.org/article/female-inmates-sterilized-in-california-prisons-without-approval/
15. H. Friedlander, *The Origins of Nazi Genocide: From Euthanasia to the Final Solution* (Chapel Hill: University of North Carolina Press, 2000), https://books.google.com/books?id=xKjqCQAAQBAJ, 12.
16. Friedlander, *The Origins of Nazi Genocide*, 126.
17. "Nazi Persecution of the Mentally & Physically Disabled," https://www.jewishvirtuallibrary.org/nazi-persecution-of-the-mentally-and-physically-disabled

18. "Nazi Persecution of the Mentally & Physically Disabled."
19. Friedlander, *The Origins of Nazi*, 39.
20. *Trials of War Criminals before the Nuremberg Military Tribunals under Control Council Law No. 10. Nuremberg, October 1946 - April 1949*, vol. 1 (Nuremberg Military Tribunals, 1949), 846 https://archive.org/details/TrialsOfWarCriminalsBeforeTheNurembergMilitaryTribunalsUnderControlCouncil.
21. Friedlander, *The Origins of Nazi Genocide*, 110.
22. *Trials of War Criminals before the Nuremberg Military Tribunals*, 67.
23. "Deadly Medicine: Irmgard Huber," https://encyclopedia.ushmm.org/content/en/article/deadly-medicine-irmgard-huber
24. Yang, *Tombstone*, 16.
25. Amartya Sen, "Democracy as a Universal Value," in *Applied Ethics*, 6th ed. (Routledge, 2017), 107–117, https://doi.org/10.4324/9781315097176, 5.

Chapter 9

1. Barry Bearak, "South Africa's President to Quit Under Pressure," *The New York Times*, September 20, 2008, sec. World, https://www.nytimes.com/2008/09/21/world/africa/21safrica.html
2. Pride Chigwedere et al., "Estimating the Lost Benefits of Antiretroviral Drug Use in South Africa," *JAIDS Journal of Acquired Immune Deficiency Syndromes* 49, no. 4 (December 1, 2008): 410–415, https://doi.org/10.1097/QAI.0b013e31818a6cd5
3. Jack Begg, "Word for Word/Nameless Dread; 20 Years Ago, the First Clues to the Birth of a Plague," *The New York Times*, June 3, 2001, sec. Week in Review, https://www.nytimes.com/2001/06/03/weekinreview/word-for-word-nameless-dread-20-years-ago-first-clues-birth-plague.html
4. "A Timeline of HIV and AIDS," HIV.gov, https://www.hiv.gov/hiv-basics/overview/history/hiv-and-aids-timeline
5. Lawrence K. Altman, "Rare Cancer Seen in 41 Homosexuals," *The New York Times*, July 3, 1981, sec. U.S., https://www.nytimes.com/1981/07/03/us/rare-cancer-seen-in-41-homosexuals.html
6. Ingrid T. Katz and Brendan Maughan-Brown, "Improved Life Expectancy of People Living with HIV: Who Is Left Behind?," *The Lancet HIV* 4, no. 8 (August 1, 2017): e324–326, https://www.thelancet.com/journals/lanhiv/article/PIIS2352-3018(17)30086-3/fulltext
7. Video: *Margaret Heckler & Robert Gallo - 1984 Press Conference*, 2013, https://www.youtube.com/watch?v=k6zd3gdDKG8. This footage does not include the vaccine comment.
8. S. C. Kalichman, *Denying AIDS: Conspiracy Theories, Pseudoscience, and Human Tragedy* (Springer New York, 2009), 35.
9. "Interviews - Margaret Heckler | The Age of Aids | FRONTLINE | PBS," https://www.pbs.org/wgbh/pages/frontline/aids/interviews/heckler.html
10. E. Kübler-Ross, *On Death and Dying: What the Dying Have to Teach Doctors, Nurses, Clergy and Their Own Families* (Abingdon, Oxfordshire: Routledge, 2009), https://books.google.com/books?id=ar2lqlxsHeQC
11. N. Nattrass, *The AIDS Conspiracy: Science Fights Back* (New York: Columbia University Press, 2013), https://books.google.com/books?id=almsAgAAQBAJ
12. Kalichman, *Denying AIDS*, 25.
13. Peter Duesberg. "HIV Is Not the Cause of AIDS." *Science*, July 29, 1988. https://doi.org/10.1126/science.3399880.
14. Kalichman, *Denying AIDS*, 38.
15. Kalichman. 25.

16. Peter H. Duesberg et al., "Withdrawn: HIV-AIDS Hypothesis Out of Touch with South African AIDS - A New Perspective," *Medical Hypotheses*, July 19, 2009, https://doi.org/10.1016/j.mehy.2009.06.024
17. Duesberg, "Withdrawn." 514.
18. "Elsevier Retracts Duesberg's AIDS Denialist Article | AIDSTruth.Org," https://aidstruth.org/news/2009/elsevier-retracts-duesberg%E2%80%99s-aids-denialist-article/
19. "Berkeley Drops Probe of Duesberg after Finding 'Insufficient Evidence,'" https://www.science.org/content/article/berkeley-drops-probe-duesberg-after-finding-insufficient-evidence
20. Kalichman, *Denying AIDS*, 56.
21. Elvira van Noort, "AEGiS-DMG: Matthias Rath's Ads 'Reckless in the Extreme,'" June 23, 2008, https://web.archive.org/web/20080623232422/http://www.aegis.org/news/dmg/2005/mg050807.html
22. Video: *Matthias Rath - The Human Cost*, 2008, https://www.theguardian.com/world/video/2008/sep/12/matthias.rath.aids.south.africa
23. C. Maggiore, *What If Everything You Thought You Knew about AIDS Was Wrong?* (American Foundation for AIDS Alternatives, 2000), https://books.google.com/books?id=pN8JAQAAMAAJ
24. Charles Orenstain and Daniel Costello, "A Mother's Denial, a Daughter's Death," *LA Times*, n.d., https://www.aidstruth.org/denialism/denialists/maggiore-latimes/
25. Orenstain, "A Mother's Denial".
26. Nattrass, *The AIDS Conspiracy*, 124.
27. Celia Farber, "Out of Control," *Harper's Magazine*, March 2006, 49.
28. "The Durban Declaration," *Nature* 406, no. 6791 (July 2000): 15-16, https://doi.org/10.1038/35017662
29. Hansen's testimony can be seen at https://youtu.be/UVz67cwmxTM
30. Andrew C. Revkin, "NASA Office Is Criticized on Climate Reports," *The New York Times*, June 3, 2008, sec. Science, https://www.nytimes.com/2008/06/03/science/earth/03nasa.html
31. M. E. Mann, *The Hockey Stick and the Climate Wars: Dispatches from the Front Lines* (New York: Columbia University Press, 2013), https://books.google.com/books?id=klerAgAAQBAJ
32. "Robert C. Gallo - Robert C. Gallo - Office of NIH History and Stetten Museum," June 7, 2020, https://web.archive.org/web/20200607235002/https://history.nih.gov/display/history/Robert+C.+Gallo
33. Anthony Brink, "HIV & AIDS - Debating AZT - The Pope of AIDS," http://www.virusmyth.com/aids/hiv/abpope.html/
34. Kalichman, *Denying AIDS*, 133.
35. Nattrass, *The AIDS Conspiracy*, 78.
36. Kalichman, *Denying AIDS*, 136-138.
37. Kalichman, 130.
38. "Please Fire Manto Now," November 28, 2006, https://web.archive.org/web/20061128154116/http://www.news24.com/News24/South_Africa/Aids_Focus/0%2C%2C2-7-659_1993915%2C00.html
39. Pride Chigwedere, George R. III Seage, Sofia Gruskin, Tun-Hou Lee, and M. Essex. "Estimating the Lost Benefits of Antiretroviral Drug Use in South Africa." *JAIDS Journal of Acquired Immune Deficiency Syndromes* 49, no. 4 (December 1, 2008): 410-415. https://doi.org/10.1097/QAI.0b013e31818a6cd5; N. Nattrass. *The AIDS Conspiracy: Science Fights Back*. (Columbia University Press, 2013). https://books.google.com/books?id=almsAgAAQBAJ, 410.

40. "South Africa: Number of Deaths from AIDS 2002-2022," Statista, https://www.statista.com/statistics/1331607/number-of-deaths-from-aids-in-south-africa/.
41. "World Development Indicators | DataBank," https://databank.worldbank.org/reports.aspx?source = 2&series = SH.DYN.AIDS.ZS&country=#
42. "HIV Rates by Country 2022," https://worldpopulationreview.com/country-rankings/hiv-rates-by-country
43. German Lopez, "A Public Health Setback," *The New York Times*, October 2, 2023, sec. Briefing, https://www.nytimes.com/2023/10/02/briefing/pepfar-aids-funding.html

Chapter 10

1. Matthew J. Belvedere, "Trump Says He Trusts China's Xi on Coronavirus and the US Has It 'Totally under Control,'" CNBC, https://www.cnbc.com/2020/01/22/trump-on-coronavirus-from-china-we-have-it-totally-under-control.html
2. C. C. Group, *Lessons from the Covid War: An Investigative Report* (PublicAffairs, 2023), https://books.google.com/books?id = gEedEAAAQBAJ/, 112, 141.
3. U.S. Department of Health and Human Services Office of the Assistant Secretary for Preparedness and Response, *Crimson Contagion 2019*, 2019, http://archive.org/details/crimson-contagion-2019/
4. "Department of Health and Human Services (HHS) Crimson Contagion 2019 Functional Exercise After-Action Report, 2020," n.d., 76.
5. David E. Sanger et al., "Before Virus Outbreak, a Cascade of Warnings Went Unheeded," *The New York Times*, March 19, 2020, sec. U.S., https://www.nytimes.com/2020/03/19/us/politics/trump-coronavirus-outbreak.html/
6. Amy Maxmen and Jeff Tollefson, "The Problem with Pandemic Planning," *Nature* 584 (August 6, 2020): 28.
7. Katie Pearce, "Pandemic Simulation Exercise Spotlights Massive Preparedness Gap," The Hub, November 6, 2018, https://hub.jhu.edu/2019/11/06/event-201-health-security/
8. Lisa Monaco, "Pandemic Disease Is a Threat to National Security," *Foreign Affairs*, March 3, 2020, https://www.foreignaffairs.com/world/pandemic-disease-threat-national-security?check_logged_in=1&utm_medium=promo_email&utm_source =lo_flows&utm_campaign= registered_user_welcome&utm_term = email_1&utm_content = 20230517
9. Sanger, "Before Virus Outbreak, a Cascade of Warnings Went Unheeded."
10. "Chinese Officials Investigate Cause of Pneumonia Outbreak in Wuhan," *Reuters*, December 31, 2019, sec. Healthcare & Pharma, https://www.reuters.com/article/us-china-health-pneumonia-idUSKBN1YZ0GP
11. Belvedere, "Trump Says He Trusts China's Xi on Coronavirus and the US Has It 'Totally under Control.'"
12. Birx, D. *Silent Invasion: The Untold Story of the Trump Administration, Covid-19, and Preventing the Next Pandemic Before It's Too Late* (New York: HarperCollins, 2022).
13. B. Woodward, *Rage* (Simon & Schuster, 2020), XIV.
14. Woodward, *Rage*, XIV.
15. Woodward, xvii.
16. Eunjung Cho, "US Officials Say Coronavirus Risk Still Low to Americans," VOA, February 7, 2020, https://www.voanews.com/a/science-health_coronavirus-outbreak_us-officials-say-coronavirus-risk-still-low-americans/6183898.html/
17. Woodward, *Rage*. xix.

18. Birx, *Silent Invasion: The Untold Story of the Trump Administration, Covid-19, and Preventing the Next Pandemic Before It's Too Late*, 56.
19. "Trump Says Coronavirus under Control in the U.S," *Reuters*, February 24, 2020, sec. Health, https://www.reuters.com/article/us-china-health-usa-trump-idINKCN20I2HV/
20. Pam Belluck and Noah Weiland, "C.D.C. Officials Warn of Coronavirus Outbreaks in the U.S.," *The New York Times*, February 25, 2020, sec. Health, https://www.nytimes.com/2020/02/25/health/coronavirus-us.html/
21. Steven Nelson, "CDC Director Says Coronavirus Outbreak Might Not Be Inevitable," https://nypost.com/2020/02/27/cdc-director-downplays-claim-that-coronavirus-spread-is-inevitable/
22. Woodward, *Rage*, 251.
23. "A Timeline of What Trump Has Said on Coronavirus - CBS News," April 3, 2020, https://www.cbsnews.com/news/timeline-president-donald-trump-changing-statements-on-coronavirus/
24. "All the Times Trump Compared Covid-19 to the Flu, Even after He Knew Covid-19 Was Far More Deadly," https://www.forbes.com/sites/tommybeer/2020/09/10/all-the-times-trump-compared-covid-19-to-the-flu-even-after-he-knew-covid-19-was-far-more-deadly/?sh = 5a0769cf9d2f
25. CBS News, "A Timeline of What Trump Has Said on Coronavirus," https://www.cbsnews.com/news/timeline-president-donald-trump-changing-statements-on-coronavirus/
26. Oliver Laughland, "Trump Tells US 'Relax, We're Doing Great' as His Virus Expert Says Worst Is Yet to Come," *The Guardian*, March 16, 2020, sec. World News, https://www.theguardian.com/world/2020/mar/15/trump-coronavirus-relax-were-doing-great-expert-worst-to-come
27. Woodward, *Rage*. xviii.
28. "Remarks by President Trump, Vice President Pence, and Members of the Coronavirus Task Force in Press Briefing – The White House," https://trumpwhitehouse.archives.gov/briefings-statements/remarks-president-trump-vice-president-pence-members-coronavirus-task-force-press-briefing-3/
29. Maggie Haberman and David E. Sanger, "Trump Says Coronavirus Cure Cannot 'Be Worse Than the Problem Itself,'" *The New York Times*, March 23, 2020, sec. U.S., https://www.nytimes.com/2020/03/23/us/politics/trump-coronavirus-restrictions.html/
30. "Restless Trump Hopes Country Can Go Back to Normal by Easter — as Health Experts Urge Caution," *People Magazine*, https://people.com/politics/donald-trump-hopes-coronavirus-shutdowns-stop-by-easter/
31. "United States COVID - Coronavirus Statistics - Worldometer," https://www.worldometers.info/coronavirus/country/us/
32. Birx, *Silent Invasion*, 152.
33. Noah Higgins-Dunn, "Trump Says There's Light at the End of the Tunnel with Coronavirus Vaccine and Treatment Research," CNBC, https://www.cnbc.com/2020/04/06/coronavirus-fight-trump-says-theres-light-at-the-end-of-the-tunnel-with-vaccine-and-treatment-research.html
34. Woodward, *Rage*, 353.
35. ABC News, "Timeline: Tracking Trump alongside Scientific Developments on Hydroxychloroquine," https://abcnews.go.com/Health/timeline-tracking-trump-alongside-scientific-developments-hydroxychloroquine/story?id = 72170553
36. ABC News, "Timeline."
37. David R. Boulware et al., "A Randomized Trial of Hydroxychloroquine as Postexposure Prophylaxis for Covid-19," *New England Journal of Medicine* 383,

no. 6 (August 6, 2020): 517–525, https://doi.org/10.1056/NEJMoa2016638, 522-523.
38. "Most Republicans Who Have Heard of Ivermectin as a COVID-19 Treatment Think It May Be Effective | YouGov," https://today.yougov.com/topics/politics/articles-reports/2021/09/02/most-republicans-who-have-heard-ivermectin
39. Birx, *Silent Invasion*, 190.

Chapter 11

1. Daniel Victor, Lew Serviss, and Azi Paybarah, "In His Own Words, Trump on the Coronavirus and Masks," *The New York Times*, October 2, 2020, sec. U.S., https://www.nytimes.com/2020/10/02/us/politics/donald-trump-masks.html/
2. The Editors of Encyclopaedia Britannica, "Influenza Pandemic of 1918–19," March 28, 2023, https://www.britannica.com/event/influenza-pandemic-of-1918-1919
3. Emily Anthes, "C.D.C. Virus Tests Were Contaminated and Poorly Designed, Agency Says," *The New York Times*, December 15, 2021, sec. Health, https://www.nytimes.com/2021/12/15/health/cdc-covid-tests-contaminated.html/
4. "'I Don't Kid': Trump Says He Wasn't Joking about Slowing Coronavirus Testing - POLITICO," https://www.politico.com/news/2020/06/23/trump-joking-slowing-coronavirus-testing-335459/
5. Birx, *Silent Invasion*, 81.
6. CDC [@CDCgov], "CDC Does Not Currently Recommend the Use of Facemasks to Help Prevent Novel #coronavirus. Take Everyday Preventive Actions, like Staying Home When You Are Sick and Washing Hands with Soap and Water, to Help Slow the Spread of Respiratory Illness. #COVID19, Https://Bit.Ly/37Ay6Cm Https://T.Co/yzWTSgt2IV," Tweet, *Twitter*, February 27, 2020, https://twitter.com/CDCgov/status/1233134710638825473
7. Birx, *Silent Invasion*, 85.
8. Birx, 84.
9. Birx, 179.
10. "United States COVID - Coronavirus Statistics – Worldometer," https://www.worldometers.info/coronavirus/country/us/
11. Victor, "In His Own Words, Trump on the Coronavirus and Masks."
12. Victor.
13. Victor.
14. Victor.
15. "The Data Is in — Stop the Panic and End the Total Isolation | The Hill," 2023, https://thehill.com/opinion/healthcare/494034-the-data-are-in-stop-the-panic-and-end-the-total-isolation/
16. Dr. Tom Frieden [@DrTomFrieden], "Herd Immunity Means 1 Million Dead Americans. That's What It Would Take to Get to Herd Immunity. That's Not a Plan—That's a Catastrophe.," Tweet, *Twitter*, August 13, 2020, https://twitter.com/DrTomFrieden/status/1293923422087651331/
17. Noah Weiland et al., "A New Coronavirus Adviser Roils the White House with Unorthodox Ideas," *The New York Times*, September 2, 2020, sec. U.S., https://www.nytimes.com/2020/09/02/us/politics/trump-scott-atlas-coronavirus.html/
18. Weiland, "A New Coronavirus Adviser Roils the White House with Unorthodox Ideas."
19. Zeynep Tufekci, "Opinion | Here's Why the Science Is Clear That Masks Work," *The New York Times*, March 10, 2023, sec. Opinion, https://www.nytimes.com/2023/03/10/opinion/masks-work-cochrane-study.html/

20. Fiona Godlee, Jane Smith, and Harvey Marcovitch, "Wakefield's Article Linking MMR Vaccine and Autism Was Fraudulent," *BMJ* 342 (January 6, 2011): c7452, https://www.bmj.com/content/342/bmj.c7452
21. Olga Khazan, "LA's Richest Neighborhoods Have Vaccination Rates Lower Than the Poorest Parts of Africa," *The Atlantic*, September 16, 2014, https://www.theatlantic.com/health/archive/2014/09/wealthy-la-schools-vaccination-rates-are-as-low-as-south-sudans/380252/
22. Gallup, "Gallup Vault: New Vaccines Not Wildly Popular in U.S.," Gallup.com, September 10, 2020, https://news.gallup.com/vault/319976/gallup-vault-new-vaccines-not-wildly-popular.aspx/
23. Gallup, "Democratic, Republican Confidence in Science Diverges," Gallup.com, July 16, 2021, https://news.gallup.com/poll/352397/democratic-republican-confidence-science-diverges.aspx/
24. "Trust in Science Is Becoming More Polarized, Survey Finds | University of Chicago News," January 28, 2022, https://news.uchicago.edu/story/trust-science-becoming-more-polarized-survey-finds/
25. "Fact Sheet: Explaining Operation Warp Speed | HHS.Gov," https://web.archive.org/web/20201219231756/https://www.hhs.gov/coronavirus/explaining-operation-warp-speed/index.html/
26. Alana Wise, "Trump Encourages His Supporters To Get COVID-19 Vaccine, Within Limits of 'Freedoms,'" *NPR*, March 16, 2021, sec. Politics, https://www.npr.org/2021/03/16/978008056/trump-encourages-his-supporters-to-get-covid-19-vaccine-within-limits-of-freedom.
27. Group, *Lessons from the Covid War*, 207.
28. Bradley J. Larsen et al., "Counter-Stereotypical Messaging and Partisan Cues: Moving the Needle on Vaccines in a Polarized United States," *Science Advances* 9, no. 29 (2023): eadg9434.
29. *Trump Encourages Vaccination* (May 2021), 2021, https://www.youtube.com/watch?v=INH-CmCgIYs.
30. "The 2021 Global Health Security Index," GHS Index, https://www.ghsindex.org/
31. "United States COVID - Coronavirus Statistics – Worldometer."

Chapter 12

1. Flora Charner, "Bolsonaro Continues to Dismiss Covid-19 Threat as Cases Skyrocket in Brazil," CNN, May 2020, https://www.cnn.com/2020/05/08/americas/brazil-coronavirus-bolsonaro-response-intl/index.html.Flora
2. "Trump Mocks Biden for Wearing Masks: 'Every Time You See Him, He's Got a Mask,'" https://news.yahoo.com/trump-mocks-biden-wearing-masks-023309905.html/
3. ABC News, "'I Had to Do It': Trump Suggests He Got Virus as Act of Political Courage," https://abcnews.go.com/Politics/trump-suggests-virus-act-political-courage/story?id = 73452023
4. Noah Weiland et al., "Trump Was Sicker Than Acknowledged With Covid-19," *The New York Times*, February 11, 2021, sec. U.S., https://www.nytimes.com/2021/02/11/us/politics/trump-coronavirus.html/
5. ABC News, "'I Had to Do It.'"
6. B. Douglas Bernheim et al., "The Effects of Large Group Meetings on the Spread of COVID-19: The Case of Trump Rallies," SSRN Scholarly Paper (Rochester, NY: Social Science Research Network, October 30, 2020), https://doi.org/10.2139/ssrn.3722299

7. Lunna Lopes et al., "KFF COVID-19 Vaccine Monitor: January 2021 - Vaccine Hesitancy," *KFF* (blog), January 27, 2021, https://www.kff.org/report-section/kff-covid-19-vaccine-monitor-january-2021-vaccine-hesitancy/
8. "Vaccines Drive Optimism about Containing COVID-19 Pandemic — CBS News Poll," https://www.cbsnews.com/news/covid-19-vaccines-optimism-pandemic-opinion-poll/
9. David R. Jones and Monika L. McDermott, "Partisanship and the Politics of COVID Vaccine Hesitancy," *Polity* 54, no. 3 (2022).
10. "United States - COVID-19 Overview - Johns Hopkins," Johns Hopkins Coronavirus Resource Center, https://coronavirus.jhu.edu/region/united-states/
11. Group, *Lessons from the Covid War*, 123.
12. "Vaccinations - Global Epidemics," https://globalepidemics.org/vaccinations/
13. Jones, "The Changing Political Geography of COVID-19 Over the Last Two Years."
14. Jones.
15. Jacob Wallace, Paul Goldsmith-Pinkham, and Jason L. Schwartz, "Excess Death Rates for Republicans and Democrats During the COVID-19 Pandemic," Working Paper, Working Paper Series (National Bureau of Economic Research, September 2022), https://doi.org/10.3386/w30512
16. "America's Culture Wars Extend into Medicine," *The Economist*, https://www.economist.com/united-states/2023/01/08/americas-culture-wars-extend-into-medicine/
17. "America's Culture Wars Extend into Medicine," *The Economist*, https://www.economist.com/united-states/2023/01/08/americas-culture-wars-extend-into-medicine/
18. "Two Years of U.S. COVID-19 Vaccines Have Prevented Millions of Hospitalizations and Deaths," December 13, 2022, https://doi.org/10.26099/whsf-fp90
19. David Wallace-Wells, "The Myth of Early Pandemic Polarization," *New York Times*, June 28, 2023, https://www.nytimes.com/2023/06/28/opinion/covid-pandemic-2020-or-covid-pandemic-politics.html?smid = nytcore-ios-share&referringSource = articleShare/
20. Gallup, "More Americans Now Willing to Get COVID-19 Vaccine," Gallup.com, November 17, 2020, https://news.gallup.com/poll/325208/americans-willing-covid-vaccine.aspx/
21. "Trump's Effort to Disavow Operation Warp Speed Shows How Far He's Fallen," https://nypost.com/2023/02/05/trumps-effort-to-disavow-operation-warp-speed-shows-how-far-hes-fallen/
22. Wikipedia contributors, "COVID-19 Pandemic Deaths — Wikipedia, The Free Encyclopedia," 2023, https://en.wikipedia.org/w/index.php?title = COVID-19_pandemic_deaths&oldid = 1132280751
23. Lucas Ferrante et al., "How Brazil's President Turned the Country into a Global Epicenter of COVID-19," *Journal of Public Health Policy* 42, no. 3 (2021): 439–451, https://www.ncbi.nlm.nih.gov/pmc/articles/PMC8393776/
24. L. B. Nucci et al., "Excess Mortality Associated with COVID-19 in Brazil: 2020–2021," *Journal of Public Health*, December 31, 2021, fdab398, https://doi.org/10.1093/pubmed/fdab398
25. Jack Nicas, "Brazilian Leader Accused of Crimes against Humanity in Pandemic Response," *The New York Times*, October 19, 2021, sec. World, https://www.nytimes.com/2021/10/19/world/americas/bolsonaro-covid-19-brazil.html/
26. Ferrante, "How Brazil's President Turned the Country into a Global Epicenter of COVID-19."
27. Flora Charner, "Bolsonaro Continues to Dismiss Covid-19 Threat as Cases Skyrocket in Brazil," CNN, May 8, 2020, https://www.cnn.com/2020/05/08/americas/brazil-coronavirus-bolsonaro-response-intl/index.html/

28. "Brazil's Bolsonaro Warns Virus Vaccine Can Turn People into 'crocodiles' | CTV News," https://www.ctvnews.ca/health/coronavirus/brazil-s-bolsonaro-warns-virus-vaccine-can-turn-people-into-crocodiles-1.5237678
29. Marcia Reverdosa, Rodrigo Pedroso, and Tara John, "Brazil's Jair Bolsonaro Tests Positive for Covid-19 after Months of Dismissing the Seriousness of the Virus," CNN, July 7, 2020, https://www.cnn.com/2020/07/07/americas/brazil-bolsonaro-positive-coronavirus-intl/index.html/
30. "Brazil Leader Rapped for Stirring Doubt on COVID-19 Vaccine | AP News," https://apnews.com/article/virus-outbreak-ap-top-news-latin-america-76e0d15bff4cf885f29abb953a842253
31. "Brazil's Bolsonaro Rejects COVID-19 Shot, Calls Masks Taboo | AP News," https://apnews.com/article/pandemics-brazil-health-coronavirus-pandemic-latin-america-0295d39d3032aa14c6675b8b4080e8cc
32. Reuters, "Election Victory, Death or Prison: Bolsonaro Names His Three Alternatives for 2022," *The Guardian*, August 29, 2021, sec. World news, https://www.theguardian.com/world/2021/aug/29/election-victory-death-or-prison-bolsonaros-names-his-three-alternatives-for-2022/

Chapter 13

1. Jeremy Schulman, "Every Insane Thing Donald Trump Has Said About Global Warming – Mother Jones," *Mother Jones*, December 12, 2018, https://www.motherjones.com/environment/2016/12/trump-climate-timeline/
2. "LBJ Presidential Library | Research," http://www.lbjlibrary.net/collections/selected-speeches/1965/02-08-1965.html
3. Panel and Tukey, *Restoring the Quality of Our Environment: Report*. https://books.google.com/books?id=mdXanDtshX8C, 126–127.
4. "Statement on Signing the Montreal Protocol on Ozone-Depleting Substances | The American Presidency Project," https://www.presidency.ucsb.edu/documents/statement-signing-the-montreal-protocol-ozone-depleting-substances/.
5. Philip Shabecoff and Special to the New York Times, "Global Warming Has Begun, Expert Tells Senate," *The New York Times*, June 24, 1988, sec. U.S., https://www.nytimes.com/1988/06/24/us/global-warming-has-begun-expert-tells-senate.html/
6. "UN Framework Convention on Climate Change – UNFCCC," IISD Earth Negotiations Bulletin, http://enb.iisd.org/negotiations/un-framework-convention-climate-change-unfccc
7. People Magazine,. "Bush at 60: 'I Really Do Feel Young.'" Accessed March 31, 2023, https://people.com/celebrity/bush-at-60-i-really-do-feel-young/.
8. *John McCain: Climate Change and the Environment*, 2007, https://www.youtube.com/watch?v=KQlX13tUSh8
9. Jane Mayer, "Koch Pledge Tied to Congressional Climate Inaction," *The New Yorker*, June 30, 2013, http://www.newyorker.com/news/news-desk/koch-pledge-tied-to-congressional-climate-inaction
10. "President Obama: The United States Formally Enters the Paris Agreement," Whitehouse.gov, September 3, 2016, https://obamawhitehouse.archives.gov/blog/2016/09/03/president-obama-united-states-formally-enters-paris-agreement/
11. Jeremy Schulman, "Every Insane Thing Donald Trump Has Said about Global Warming – Mother Jones," *Mother Jones*, December 12, 2018, https://www.motherjones.com/environment/2016/12/trump-climate-timeline/
12. Schulman, "Every Insane Thing Donald Trump Has Said about Global Warming."
13. "Statement by President Trump on the Paris Climate Accord – The White House," https://trumpwhitehouse.archives.gov/briefings-statements/statement-president-trump-paris-climate-accord/
14. Joravsky, *The Lysenko Affair*, 94.

15. *President Trump 2018 State of the Union Address (C-SPAN)*, 2018, https://www.youtube.com/watch?v = ATFwMO9CebA. His comments on coal begin at 42:04 in the video.
16. Kelsey Tamborrino, "Big Costs, Sweeping Changes: What to Know about the IRA," POLITICO, August 16, 2023, https://www.politico.com/news/2023/08/16/democrats-climate-law-listicle-00111360
17. Brad Plumer, "In a First, Renewable Energy Is Poised to Eclipse Coal in U.S.," *The New York Times*, May 13, 2020, sec. Climate, https://www.nytimes.com/2020/05/13/climate/coronavirus-coal-electricity-renewables.html
18. N. Oreskes and E. M. Conway, *Merchants of Doubt: How a Handful of Scientists Obscured the Truth on Issues from Tobacco Smoke to Global Warming* (Bloomsbury USA, 2010), https://books.google.com/books?id = MGq0wAEACAAJ. This book is essential reading on the role of corporations in denying climate science.
19. Exxon merged with Mobil in 1999 to become ExxonMobil.
20. Shannon Hall, "Exxon Knew about Climate Change Almost 40 Years Ago," *Scientific American*, https://www.scientificamerican.com/article/exxon-knew-about-climate-change-almost-40-years-ago/
21. Oreskes *Merchants of Doubt*; J. L. Powell, *The Inquisition of Climate Science*, Mersion: Emergent Village Resources for Communities of Faith Series (New York: Columbia University Press, 2011), https://books.google.com/books?id = YKaOOcWyNywC
22. From Kathy Mulvey, Seth Shulman, Dave Anderson, Nancy Cole, Jayne Piepenburg, and Jean Sideris. "The Climate Deception Dossiers: Internal Fossil Fuel Industry Memos Reveal Decades of Corporate Disinformation." *Union of Concerned Scientists*, 2015, 9.
23. Mulvey, "The Climate Deception Dossiers."
24. G. Supran, S. Rahmstorf, and N. Oreskes, "Assessing ExxonMobil's Global Warming Projections," *Science* 379, no. 6628 (January 13, 2023): eabk0063, https://doi.org/10.1126/science.abk0063
25. Supran.
26. Supran, Rahmstorf, and Oreskes.
27. Brendan DeMelle, "ExxonMobil's Funding of Climate Science Denial," *DeSmog* (blog), https://www.desmog.com/exxonmobil-funding-climate-science-denial/
28. IPCC, "Summary for Policymakers," in *Climate Change 2022: Mitigation of Climate Change. Contribution of Working Group III to the Sixth Assessment Report of the Intergovernmental Panel on Climate Change*, ed. P. R. Shukla et al. (Cambridge, UK, and New York, NY, USA: Cambridge University Press, 2022), https://doi.org/10.1017/9781009157926.001
29. "IPCC AR6 Summary for Policymakers" (IPCC, March 14, 2023), https://www.ipcc.ch/report/ar6/syr/
30. Elizabeth Kolbert, "Why Did the Biden Administration Approve the Willow Project?," *The New Yorker*, March 13, 2023, https://www.newyorker.com/news/daily-comment/why-did-the-biden-administration-approve-the-willow-project
31. *Biden: "No Ability For The Oil Industry To Continue To Drill, Period, Ends,"* 2020, https://www.youtube.com/watch?v = viAXGth3gQA
32. Oliver Milman, "Biden's Approval of Willow Project Shows Inconsistency of US's First 'Climate President,'" *The Guardian*, March 14, 2023, sec. Environment, https://www.theguardian.com/environment/2023/mar/14/biden-president-approved-alaska-willow-project
33. Ben Lefebvre, "Biden Administration Approves Controversial Alaskan Oil Project - POLITICO," March 13, 2023, https://www.politico.com/news/2023/03/13/biden-administration-approved-willow-alaska-oil-00086746
34. Ryan Lizza, Rachel Bade, and Eugene Daniels, "Playbook: Biden's Crude Move to the Center - POLITICO," March 13, 2023, https://www.politico.com/newsletters/playbook/2023/03/11/bidens-crude-move-to-the-center-00086653

35. Oliver Milman, "Al Gore Warns It Would Be 'Recklessly Irresponsible' to Allow Alaska Oil Drilling Plan," *The Guardian*, March 10, 2023, sec. US news, https://www.theguardian.com/us-news/2023/mar/10/al-gore-biden-alaska-oil-drilling-willow-development
36. Max Bearak, "It's Not Just Willow: Oil and Gas Projects Are Back in a Big Way," *The New York Times*, April 6, 2023, sec. Climate, https://www.nytimes.com/2023/04/06/climate/oil-gas-drilling-investment-worldwide-willow.html
37. Kolbert, "Why Did the Biden Administration Approve the Willow Project?"

Chapter 14

1. Lisa Friedman, "A Republican 2024 Climate Strategy: More Drilling, Less Clean Energy, *New York Times*, August 4, 2023, https://www.nytimes.com/2023/08/04/climate/republicans-climate-project2025.html

Index

For the benefit of digital users, indexed terms that span two pages (e.g., 52–53) may, on occasion, appear on only one of those pages.

abortions, 119–20
Adams, Jerome, 135
aether, 84
African National Congress, 116
agronomy, 11, 43, 70–72
AIDS
   death toll, 107, 119
   denialism, 110
   false remedies, 112, 113, 117–19
   origins of, 107
   and pseudoscience, 4
   research on, 107–10
   scientific consensus, 114–15
   South African response, 116–19
   symptoms, 107–8
Alaska, oil drilling, 165
All-Union Institute for Plant Breeding and Genetics (AIPB), 38–39
All-Union Society for Cultural Relations with Foreign Countries, 58–59
American Breeder's Association, 93
American Clean Energy and Security Act, 157
American Enterprise Institute, 164
American Petroleum Institute, 162–63
Americans for Prosperity, 158
antisemitism, 76, 78
   and eugenics, 95
   in science, 77, 81–82, 85–87, 89
anti-vax movement, 138, 141, 150
apartheid, 116
Arctic, rising temperatures, 154, 166
Arrhenius, Svante, 154, 162–63
Aryan physics movement, 86
Aryan supremacy, 4, 78, 91
Association of German Natural Scientists for the Preservation of Pure Science, 84

asymptomatic spread, 133, 135
Atlas, Scott, 136–37, 145
atomic bomb, 76–77, 79–80
autism, false link to vaccines, 138–39
Azar, Alex, 125–27, 135
AZT, 111

Barulina, Yelena, 41–42
Bateson, William, 14
Belgium, 80–81, 102
Bell, Alexander Graham, 93
Bengtsson, Bengt, 45–46
Beria, Lavrentiy, 37
Beyerchen, Alan, 79, 86
biblical creationism, 3
Biden, Joe, 143, 160, 165
birth control, 93
Birx, Deborah, 124–25, 128, 129, 134–35, 136, 137, 139–40
Black, James, 162–63
Blair, Tony, 157
Bolsonaro, Jair, 4–5, 124, 143, 151
Bolton, John, 123
Borinskaya, Svetlana, 47
Born, Max, 76–77
Bossert, Tom, 123
Bouhler, Phillip, 98
boycotts, 78
Bragg, William, 80–81
Brandt, Karl, 98
Brazil
   COVID-19 response, 4–5, 124, 143, 151
   elections, 153
Brink, Anthony, 115–16, 117
Brown Shirts, 78
Bryan, William, 131

Buck, Carrie, 93–94
Bukharin, Nikolai, 35
Bulgaria, 63–64
Burbank, Luther, 21–22, 93
Bush, George H. W., 156
Bush, George W., 119–20, 157

cancer research, 110–11
cannibalism, 24, 74
cap-and-trade, 157–58
carbon dioxide, 154, 155–56, 162–63
cathode rays, 81–82
cell structure, 9–10
Center for Disease Control (CDC)
   and COVID-19, 123–24, 125–27, 128,
      132, 133–34, 135–36, 138
   and HIV/AIDS, 107–8
center of origin, 16–17
Chetverikov, Sergei, 9–11, 13–14
child mortality, 1
China
   communism, 27–28, 57, 58–59, 74
   Communist Party, 57, 64–65
   constitution for agriculture, 65
   drought, 57
   famine, 57
   Ministry of Higher Education, 64–65
   and pandemics, 121, 122
   Paris Agreement, 158–59
   Propaganda Department, 64
   state science policy, 3–4
   treaty with Soviet Union, 58
Chinese Academy of Sciences, 61, 62–63, 64
Chinese Communist Party (CCP), 57, 64–65
Chirac, Jacques, 109
civil servants, 78–79
climate change. *See* global warming
Clinton, Bill, 156–57
coal, 159–60
cold fusion, 167–68
collectivization of farms, 15, 23–24
communism
   Chinese, 27–28, 57, 58–59, 74
   and Jews, 79
   Soviet, 10, 19, 24, 58–59
Communist Party (China), 57, 64–65

Communist Party (USSR), 10, 11, 19–20, 45–46, 51–52
concentration camps, 64, 90–91, 95–96, 102
conditioned reflex, 9–10
ConocoPhillips, 165, 166
contact tracing, 133
cotton, 33
COVID-19
   anti-shutdown protests, 129–30
   beginning of, 4–5, 122, 123–24
   death toll, 142, 143–44, 145–47, 146f, 151
   source, 123–24
   testing, 133
   vaccines, 138, 140
creationism, 3
cremation, 99, 101
Crick, Francis, 9–10
crimes against humanity, 100–1
crimson contagion exercise, 121, 125, 128
Crookes, William, 80–81
crossbreeding, 13–14, 26
cultropreneurs, 110, 113, 138–39
cytology, 44–45, 46
Czechoslovakia, 63–64

Darwin, Charles, 3, 13, 14, 37, 92
da Silva, Luiz Inâcio, 153
death toll
   AIDS, 107, 112, 119
   COVID-19, 129, 142, 143–44, 145–47, 146f, 151
   famine, 24, 54–55, 72, 74–75, 103
   flu, 122–23, 133
   global warming, 155
   holocaust, 76, 90–91
degenerative diseases (in agronomy), 33–35
de Klerk, F. W., 116
democracies
   and science denial, 103
   and truth, 168
dense planting, 49–50, 68
Dikötter, Frank, 75
Directorate of Global Health Security and Biodefense, 122–23
DNA molecule, 9–10

Dobzhansky, Theodosius, 22, 60–61
dominant traits, 13–14
Doyarenko, Alexei, 19–20
drought
    China, 57
    Soviet Union, 49, 50f
drug use, and AIDS, 111
Duesberg, Peter, 110–13

Earth Summit (1992), 156–57
Ebola, 122–23
echo chamber effect, 167, 169
Eddington, Sir Arthur, 84
Einstein, Albert, 78, 84, 85–86, 87
Eisenhower, Dwight, 155
election campaign
    Bolsonaro, 153
    Trump, 134, 143–44
England, World War I, 80–81, 82–83
Environmental Protection Agency, 158–59, 160
eugenics, 4, 91, 95
    activism against, 99–100
European Union, carbon strategy, 157
Event 201, 123
evolution, 3
Ewald, P. P., 81
excess deaths, concept of, 119
ExxonMobile, 162, 163f, 164

Famine
    China, 57, 72, 103
    death toll, 24, 54–55, 72, 74–75, 103
    as genocide, 24
    planned, 91
    Soviet, 18–19, 23–24, 32–33
    Ukraine, 12–13, 23, 55
Farber, Celia, 114–15
farms, collectivization, 15, 23–24
farm tools, 69
Fauci, Anthony, 126, 127, 128, 129, 137, 139–40
Federal Office of Research Integrity, 109
fertilization, 68
field management, 69
Finland, 40–41
First World War. See World War I
Fleischmann, Martin, 167–68

Fleming, Alexander, 80–81
flu
    death toll, 122–23, 133
    vaccination, 147, 148f
fracking, 165–66
France, 80–81
Frieden, Tom, 136–37

Gallo, Robert, 107–9, 109f, 115
Galton, Francis, 92
gas chambers, 99, 101
gay community, and AIDS epidemic, 107
Geffen, Nathan, 112–13
general theory of relativity, 84–85
genetics
    banning of, 46–47, 60–61
    as field of study, 14, 30–31
    International Congress of, 16, 22–23, 45–46
genocide, 24, 151. See also holocaust
genotypes, 37–38
German Society for Racial Hygiene, 92
Germany
    activism against murders, 99–100
    end of World War II, 90–91
    under the Nazis, 76
    nuclear program, 86
    World War I, 80–81
    See also Nazism
Gestapo, 87
Global Climate Coalition, 162–63
Global Health Survey, 142
global warming
    denial, 3, 5
    politicization of, 155
    science of, 154, 155–56, 162–63, 163f
    villain scientist, 115
Gorbunov, Nikolai, 35
Gore, Al, 165–66
Gorky, Maxim, 10–11
Gosney, Eugene S., 95
Gould, Stephen Jay, 47
Graham, Loren, 53
Great Break, 15
Great Chinese Famine, 57
Great Leap Forward, 67, 70–72
greenhouse effect, 156

Guterres, António, 164
Guyana, 164

H1N1 virus, 122–23, 133
Haber, Fritz, 78–79
Hadamar psychiatric hospital, 90–91, 99, 100f, 100, 102f
Haeckel, Ernst, 92
Hagemann, Rudolf, 63–64
Hahn, Otto, 79–80
handicapped people, and holocaust, 97f, 98–99
Hansen, James, 115, 156
Harris, Kamala, 160
Head of State, role in science denial, 115
Heckler, Margaret, 108–9, 109f
Heisenberg, Werner, 79–80, 86, 88f
herd immunity, 136–37
hereditary defects
   cost of, 97f, 98
   as justification for sterilization, 95–97
Hereditary Health Courts, 96–97
heredity, 9–10, 13, 27, 30
Heritage Foundation, 169
hero scientists, 110–11
Hertz, Gustav, 79
Hertz, Heinrich, 81–82, 85–86
Hilfrich, Antonius, 99–100
Himmler, Heinrich, 87
Hippocrates, 13
Hitler, Adolf, 4
   death of, 98
   and eugenics movement, 95–96, 98
   ideology, 93
   imprisonment, 85
   Mein Kampf, 76, 85, 92–93, 102
   planning for war, 98
   rise to power, 78
   science advisor, 86
   view of Jewish scientists, 77, 79
HIV. *See* AIDS
Hobbes, Thomas, 2
Holmes, Oliver Wendell, 93–94
Holocaust, 4, 76, 101–2
   "euthanasia", 98
   gas chambers, 99, 101
   German activism against, 99–100
   and handicapped people, 97f, 98–99

Holodomor, 23, 55
House of Shutters, 90–91
Huber, Irmgard, 101
Hungary, 63–64
Hussein, Saddam, 168
Huxley, Julian, 14
hybridization, 63
hydroxychloroquine, 130–31, 151, 152

infant mortality, 117–19
Inflation Reduction Act, 160
inheritance of acquired characteristics, 3–4, 12–13, 27, 37, 45–46, 65–66
Inslee, Jay, 139
Intergovernmental Panel on Climate Change, 164–65
International Astronomical Union, 86
internet, as echo chamber, 167, 169
Iraq, 168
irrigation, 70
Ivanovich, Nikolai, 39
ivermectin, 152

Jaworski, Leon, 100–1
Jews
   Nazi definition of, 78–79
   as scientists, 77, 78–79, 87
   stereotypes of, 85–86, 93
   See also antisemitism, Holocaust
Johannsen, Wilhelm, 37–38
Johns Hopkins University, 76–77, 123
Johnson, Lyndon, 155
Joravsky, David, 18–20, 26–27, 36, 49–50
journals, 2

Kaiser Wilhelm Institutes, 78–79, 81, 95–96
Kaiser Wilhelm Society, 79, 81
Kaishek, Chiang, 57–58
Kalichman, Seth, 110–11, 112–13
Karpechenko, Georgy, 22, 28–29
Kelly, John F., 123
Khrushchev, Nikita, 24, 35–36, 51
   "Secret Speech", 51–52, 64
Koch brothers, 158
Kol, Alexander, 21–22
Kolbert, Elizabeth, 166
kolkhozes, 15

INDEX   193

Koltsov, Nikolai, 9–11, 35
Korean War, 58
Kubler-Ross, Elizabeth, 110
Kuomintang, 57
Kushner, Jared, 136
Kyoto Protocol, 156–57

Lamarck, Jean Baptiste, 13, 14
Lamb, Horace, 80–81
Latvia, 40–41
Lemkin, Rafael, 24
Lenard, Philipp, 81, 83*f*, 86
Lenin Academy, 25–26, 28, 29, 36–37
   conference (1948), 45
   and Lysenko, 26, 29, 32–33, 36, 38, 52, 60–61, 63, 64
   and Muller, 23, 29
   and Muralov, 35
   and Vavilov, 15–16, 17, 19, 25, 29, 39–40
Lenin, Vladimir
   and Lenin Academy, 15–16
   ideology, 10, 23–24, 27, 62
Leningrad Conference (1929), 17
Lenz, Fritz, 95–96
Lepeshinskaia, Olga, 44–45, 46
life expectancy, 1
Lithuania, 40–41
Liu Shaoqi, 61, 74–75
living icons, 110, 113–14
Lodge, Oliver, 80–81
Lubyanka prison, 41
Lu Dingyi, 64
Lula (Luiz Inácio da Silva), 153
Luo Tianyu, 60–61
Lysenko, Denis, 18–19
Lysenkoism
   in China, 3–4, 56, 58
   cost of, 54–56
   and Michurinism, 26
   in USSR, 3–4, 25–26, 30, 40, 46–49, 56
   See also Michurinism
Lysenko, Trofim, 3–4, 11–13, 12*f*, 14–15, 17
   criticism of, 17–18, 25, 29–30, 45–46, 53
   on cytology, 44–45
   death, 53

   failed experiments, 32, 44–45, 49–51
   on fertilization, 68
   on genetics, 29–30, 43
   and Khrushchev, 51
   media support, 114–15
   as president of Lenin Academy, 36, 38, 46, 48*f*
   removal from presidency of Lenin Academy, 63, 64
   rivalry with Vavilov, 15–16, 17, 22–23, 25, 28, 29–30, 37–39
   on socialist science, 26, 37–38, 39
   state support, 26, 28–29, 32, 36–37, 39–40, 46–49, 51*f*
   successes, 43–44
   support for Williams, 66–67
   vernalization, 12–13, 17–19, 22–23

Madlala-Routledge, Nozizwe, 117–19
Maggiore, Christine, 113–14
Maksimov, Nikolai, 17–18, 25, 48–49
Mallory, Walter, 57
Manchin, Joe, 160
Mandela, Nelson, 116–17
Manhattan Project, 76–77, 79–80
Mann, Michael, 115
Mao Zedong, 3–4, 56, 57
   constitution for agriculture, 65
   false shows for the benefit of, 68, 70–72
   influence of Williams, 66–67, 69
   inspection tour, 73*f*
   marriage, 91–92, 93–94
masks
   Bolsonaro's view, 152–53
   COVID-19, 134
   Donald Trump's view, 132, 135–36
   and Spanish flu, 132–33
Mason, Jeff, 136
Mbeki, Thabo, 4, 107, 112, 116–17
McCain, John, 157
McCarthy, Jenny, 138–39
Meadows, Mark, 143
measles, 139
Medvedev, Zhores, 29–30, 45
Mein Kampf, 76, 85, 92–93, 102
meiosis, 44–45
Meitner, Lise, 79–80
Mendel, Gregor, 13–14

Mendelism, 37, 39, 46
Messonnier, Nancy, 126–27
Michelson-Morley experiment, 84
Michurin, Ivan Vladimirovich, 26
Michurinism, 26
  in China, 58, 65–66
  See also Lysenkoism
Michurin Study Society, 60–61
millet, 43
Moderna, 140–41
modern synthesis, 14
Molotov, Vyacheslav, 10, 28
Molotov-Ribbentrop secret protocol, 41
Monaco, Lisa, 123
Montagnier, Luc, 107–8, 109
Montreal Protocol, 155–56
Morgan, Thomas Hunt, 21–22, 46, 60–61
Morganism, 64–65
Muller, H. J., 23, 29, 45–46
Müller, Wilhelm, 88–89
Muralov, Alexander, 35, 41–42

Nattrass, Nicoli, 110, 117
natural selection, 44–45
Nazism
  cost to science, 89
  defeat of, 45–46
  eugenics, 4
  gas chambers, 99, 101
  overhead, 80
  racial ideology, 91, 95
  See also Hitler, Adolf
Nobel Prize
  Haber, 78–79
  Heisenberg, 87
  Hitler's disapproval of, 87
  Lenard, 81–82
  Montagnier, 109
  Morgan, 46
  Muller, 23
  Pavlov, 9–10
  Planck, 79
  Sen, 103
  Watson and Crick, 9–10
nuclear energy, 157, 167–68
nuclear science, 79–80, 86
Nuremberg Laws, 95–96
Nuremberg trials, 100–1

Nuzhdin, Nikolai, 53

oak trees, 49–50
Obama, Barack, 157, 158–60
O'Brien, Robert, 125–26
Odessa Institute, 33, 38
OGPU, 21–22, 23
Operation Warp Speed, 140, 150
Oppenheimer, Robert, 76–77
Order of Lenin, 14–15, 29
ozone, 155–56

pandemic
  future, 150
  simulations, 121
  See also COVID-19, flu
pangenesis, 13
Paris Agreement, 158–59
  U.S. withdrawal, 159, 160
partisanship, 64–65
  and COVID response, 148
  distrust of medical professionals, 147
  and science denial, 139–40, 143, 169
  vaccine hesitancy, 144, 147, 148f, 149
Pauli, Wolfgang, 76–77
Pavlov, Ivan, 9–10
Peasant Labor Party (TKP), 21–22
peer review, 167–68
Pence, Mike, 126–27, 144
People's Republic of China. See China
permafrost, 166
Perry, Rick, 123
personal-belief exemption for vaccines, 139
personal protective equipment (PPE), 122, 134
pest control, 70, 71f
phenotypes, 37–38
photoelectric effect, 81–82
Planck, Max, 79, 81
ploughing, 68
Poland, 40–41
polio, 139–40
political affiliation
  and COVID response, 148
  distrust of medical professionals, 147
  and science denial, 139–40, 143, 169
  vaccine hesitancy, 144, 147, 148f, 149

Pons, Stanley, 167–68
Popenoe, Paul, 95
population genetics, 9–10
potatoes, 33–35, 43–44
Pottinger, Matt, 125
poverty, as explanation for AIDS, 116–17
praise singers, 110, 114–15
Pravda, 11
presidential debates, 143
President's Emergency Plan for AIDS Relief (PEPFAR), 119–20
Prezent, Isaak, 27–28, 47–48
Protests
  anti-eugenics, 99–100
  anti-lockdown, 129–30, 152
Pruitt, Scott, 158–59
pseudoscience, nature of, 3–4, 56
public-private partnerships, 141
public service announcements, 141–42

Qingdao Symposium, 64, 65–66
quality of life, 1–2

racial superiority, ideology of, 91
Ramsay, William, 80–81
Rasnick, David, 112, 113, 117
Rath, Matthias, 113
Reagan, Ronald, 109, 155–56, 161
recapitulation theory, 92
recessive traits, 9–10, 13–14
Redfield, Robert, 125–27
reforestation, 49
Rockefeller Foundation, 95–96
Röntgen, Wilhelm, 81–82
Rumania, 63–64
Russia
  agriculture, 11, 23–24, 26–27, 34
  revolution, 10
  scientists, 9–10, 22, 25, 53–54, 55–56
  and Ukraine, 168
  See also Soviet Union

Sakharov, Andrei, 53
Sanger, Margaret, 93
SARS virus, 123–24
Schneider, Lawrence, 59–60
school closings, 149
science, 2, 167–69

science denial, nature of, 2–4, 167
  cost of, 169
Science Society of China, 61
Scientific American, 111–12
scientific journals, 80–81, 167–68
Scovill, Eliza Jane, 114
Secret Speech, 51–52, 64
Sen, Amartya, 24, 103
shelterbelts, 49, 50*f*
Shostakovich, Dmitri, 50
Sino-Soviet Friendship Association, 58–59, 60–61, 63
Snyder, Timothy, 54–55, 91
soft inheritance, 13, 14
Sommerfeld, Arnold, 87
South Africa, 4, 107–8, 112, 113, 116–19
Soviet Union
  Academy of Sciences, 17, 25, 53–54
  agriculture, 11, 19, 23–24, 32, 43
  census, 54–55
  communism, 10, 19, 24, 58–59
  famine, 18–19, 23–24, 32–33
  Great Break, 15
  ideology, 14–15, 19, 56
  influence on Africa, 116
  influence on China, 58, 64, 65–67
  newspapers, 11, 15–16, 27, 36–37, 39, 40–41, 49, 114–15
  punishment of dissent, 10–11, 19–20, 22–23
  reforestation, 49
  state science policy, 3–4, 14, 26, 30, 40, 46–48, 56
  Supreme Soviet, 40–41
  World War II, 43, 91, 100–1
  See also Stalin, Josef
Soyfer, Valery, 40–41, 45
Spanish flu, 121, 132
sparrows, 70
spontaneous generation, 44–45
Stalin Constitution, 40
Stalin, Josef
  and China, 57–58
  criticism of, 24, 51–52, 64
  death of, 41–42, 51–52
  as general secretary of the Communist Party, 10
  Great Break, 15

Stalin, Josef (*cont.*)
  and inheritance of acquired
    characteristics, 65–66
  reforestation plan, 49, 50f
  response to Vavilov, 25, 28, 39–40
  and state science policy, 3–4, 14, 30–31,
    36–37, 46, 54–55, 56, 159
  support for Lysenko, 21, 26, 28–29, 32,
    46–47, 51f, 51–52
  and Supreme Soviet, 40
  terror, 10–11, 25, 35–36, 53–54
Stalin Prize, 14–15, 44–45, 50
Stark, Johannes, 85
sterilization, forced, 91–92, 93–95, 99
Strassmann, Fritz, 79–80
Stubbe, Hans, 63–64
sugar beets, 44
Sun Yat-Sen, 57–58
Supreme Soviet, 40
Sweden, 136–37
swine flu, 122–23, 133
Szilard, Leo, 79–80

Tan, C. C., 65
terror, Stalin's, 10–11, 25, 35–36, 53–54
testing, for COVID-19, 133
Thomson, J. J., 80–82
Tillerson, Rex, 123, 163–64
tobacco, 112
tools, farm, 69
torture, in Soviet Union, 41, 48–49
totalitarianism, 103
tractors, 69
Treaty of Peace, Security, and Friendship, 58, 59f
Treaty of Versailles, 82–83
trees, reforestation, 49, 50f
trials (legal)
  Hadamar, 90–91, 100
  Nuremberg, 90–91, 100–1
  in Soviet Union, 21–22, 36, 41
trials (scientific)
  on advertising, 142
  agricultural, 22–23, 34
  pharmaceutical, 113
Trump, Donald, 4–5
  and Biden, 153
  cabinet appointments, 117–19
  campaigning, 143–44
  on fossil fuels, 165
  on global warming, 154, 158–60
  on masks, 132, 135–36, 143
  politicization of science, 143
  press briefings, 130, 131, 137
  response to COVID-19, 124, 128, 130,
    132, 142, 143–44
  on testing, 134
  and vaccination, 141–42, 150
Trump, Melania, 141, 143
truth, importance of, 168–69
Tshabalala-Msimang, Manto, 117–19
Tsitsin, N. V., 63
Tunlid, Anna, 45–46
Tweets
  CDC, 135
  Donald Trump, 126–27, 128–30, 135–
    36, 143, 150, 158–59
  Jerome Adams, 135

Ukraine
  annexation of, 40–41
  famine, 12–13, 23, 54–55, 91
  farming, 66–67
  reforestation, 49
  in twenty-first century, 168
  World War II, 102
uncertainty principle, 86–87
United States
  agriculture, 49
  AIDS deniers, 115
  climate response, 117–19,
    155, 164–66
  COVID-19 response, 4–5, 124, 128,
    130, 133, 134, 138
  eugenics, 93–96
  pandemic preparedness, 122–23
  partisanship, 64–65, 144, 147, 148
  prosecution of war crimes, 100–1
  scientific progress, 89, 107–8, 140
  See also Trump, Donald
USSR. *See* Soviet Union

vaccination
  compulsory, 94
  and political affiliation, 144
  statistics, 139–40, 139t

vaccines, 1
  Bolsonaro's view, 152–53
  COVID-19, 138, 140
  development of, 140
  efficacy, 148
  flu, 147, 148*f*
  price of, 152
Vavilov, Nikolai, 15, 30–31, 32, 38
  death of, 41, 43
  and Lysenko, 17, 22–23, 25, 26, 28, 29–30, 37–39, 53–54
  mugshot, 42*f*
  as president of Lenin Academy, 17, 19, 28, 36, 45–46
  as target of Soviet State, 10–11, 21–22, 23, 25–26, 35, 37, 39–40
vernalization, 12–13, 17, 18–19, 32–33
victory gardens, 43–44
villain scientist, 115
  and AIDS denial, 115–16
  and global warming, 115, 156
von Braun, Wernher, 89

Wagner, Gerhard, 98
Wakefield, Andrew, 138–39
Wallace, Chris, 136
Wallace-Wells, David, 148–49
war crimes, 100
Watson, James, 9–10
Wehrmacht, 43–44
Weyland, Paul, 84
wheat, 12–13, 18–19, 23–24, 29–30, 32–33, 43–44
Whitmer, Gretchen, 129–30
Williams, Vasily, 66–67
Willow Project, 165–66
women in science, 85–86
Woodward, Robert, 125–26, 127
World Health Organization (WHO), 123–24, 138
World Trade Centre, 168
World War I, 80–81, 93
World War II, 16, 43, 89
  end of, 90–91
  preparations for, 98
  See also Holocaust
wreckers, 19

Xinhua News Agency, 74
x-rays, 81–82

Yakovlev, Yakov Arkadevich, 18–19, 26, 35–36
Yakushkin, Ivan, 21–22
Yang, Jinsheng, 102–3
Yang Jisheng, 72
Yizi, Chen, 75
Yu Guangyuan, 64–65